THE COLLEGE OF
**PHLEBOLOGY**™
PART OF THE COLLEGE OF PHLEBOLOGY SERIES

# Understanding Venous Reflux

the cause of varicose veins
and venous leg ulcers

by Professor Mark S Whiteley MS FRCS (Gen)
Visiting Professor University of Surrey

Whiteley Publishing

Published by Whiteley Publishing Ltd
Second soft cover edition 2013

ISBN 978-1-908586-00-1

# contents

# Introduction

This book has been written for two groups of people: firstly for healthcare professionals, who either treat people with venous problems or wish to do so; and secondly for members of the general public want to understand more about the cause of varicose veins, thread veins of the legs and other venous conditions such as venous leg ulcers. After all, these conditions will affect almost half the population at some stage in their lives and currently is treated by a variety of people in a variety of ways, with very different outcomes. In order to make sure they or their loved ones get the best treatment, a good basic understanding of the problem by the health care professional is essential and the ability to ask informed questions can often help when making sensible choices as to treatment options.

Although one would expect qualified people to have a thorough understanding of their subject, research in varicose veins and venous reflux disease (or venous incompetence) has progressed at a rapid rate over the last decade or so. This, coupled with the unfortunate and incorrect view of a great many people both inside and outside healthcare that varicose veins are "only cosmetic", has meant that there is very little pressure on the majority of people who treat veins to keep up with the research and our new understanding of venous physiology.

Since 1999, I have been running courses to teach healthcare professionals how to treat varicose veins, thread veins and other venous conditions such as venous eczema and leg ulcers. During that time, it has amazed me that none of the professionals coming on my courses - whether surgeons, other doctors, nurses, beauty therapists or dentists - have been able to answer all of the simple questions that I pose about how veins work and the consequences of the valves failing in different veins. Although they all know that the basic problem is that the "valves aren't working", the problems that result from this, and how they affect the legs, seem something of a mystery to them. As understanding this is essential to good venous treatment, this is clearly very worrying.

I have written this book in a simple style, that should be easy to follow, much in the way that I teach on my courses. My understanding of varicose veins and venous reflux disorders has been amalgamated from my own research and experience, coupled with a large number of other influences learned from reading research papers and listening to research presentations from other research units. I have not attempted to reference any of these, as to try and find all of the influences that have contributed to my understanding of this subject over the last 20 years would be impossible.

I hope that this resulting book will take the reader through this fascinating subject and give them a deeper understanding of this common problem. If by doing so this book helps healthcare professionals to understand their subject deeper and therefore review their treatment strategies, and helps members of the general public to understand what might be going on in their own legs or the legs of their friends and relatives in order that they may ask relevant questions of proposed treatments, then it has served its purpose.

# CHAPTER 1

# The venous system of the lower limb

## Why do we have veins in the legs?

Simply, we need to transport blood back from the tissues in the foot and leg to the heart.

Arterial blood has taken oxygen and nutrients to the living tissues of the foot and the leg where they have been used resulting in carbon dioxide and waste products which need to be taken away. These are removed in the venous blood and it is the function of the venous system of the leg to achieve this.

The systemic★ circulation related to the legs (**Figure 1**).

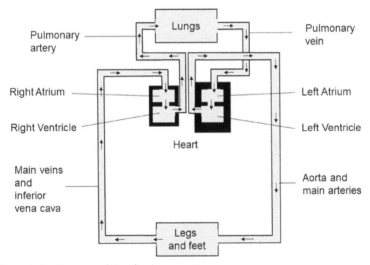

Figure 1: Basic layout of the circulatory system

*★ The 'Systemic' circulation is the circulation of blood throughout most of the body. There are some special areas of the body where the circulation differs – usually in terms of flow, pressure or structure of the blood vessels. Examples of these are the circulation through the gut (the 'portal' circulation) and circulation through the lungs (the 'pulmonary' circulation). However these do not need to be understood with regards venous reflux in the legs.★*

In the normal systemic circulation system, the heart pumps oxygenated blood, full of nutrients, from the left ventricle into the aorta and main arteries. The heart contracts (systole), sending a large volume of blood into these arteries at high pressure. To withstand this pressure, the main arteries (the aorta and its major branches, the iliac arteries) have a large amount of elastic protein in their walls called elastin. This allows the large arteries to dilate under pressure and then recoil when the heart relaxes (diastole).

There are no valves in the arterial system with the exception of the heart valves. When the left ventricle contracts in systole, the aortic valve which lies between the left ventricle and the aorta opens, allowing blood to flow out into the aorta under high pressure. When the left ventricle relaxes in diastole, the pressure of the blood in the aorta becomes higher than that in the expanding left ventricle and blood tries to re-enter the heart. At this point, the reverse of flow closes the aortic valve stopping blood from refluxing back into the left ventricle.

As the elastic tissue in the aorta and main vessels recoils, it increases pressure on the blood, forcing it to flow away from the heart and into the small arteries and arterioles.

From the arterioles, blood passes into the capillaries. It is in the capillaries where oxygen and nutrients diffuse out of the arterial blood and into the tissues, reaching the living cells that need them. Once these cells metabolise the oxygen and use the nutrients, the waste products of metabolism, carbon dioxide and other waste

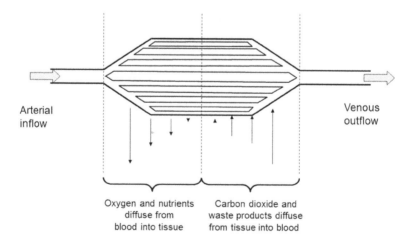

Arterial inflow

Venous outflow

Oxygen and nutrients diffuse from blood into tissue

Carbon dioxide and waste products diffuse from tissue into blood

Figure 2: Basic structure and function of capillaries in legs

products, diffuse into the venous blood (**Figure 2**).

Of course the blood is the same basic substance going from the arterial end to the venous end of the capillaries – it is merely called 'arterial blood' when it is in the arterial end of the capillaries and full of oxygen and nutrients and 'venous blood' when it has moved into the venous end of the capillaries and is full of carbon dioxide and waste products.

It is widely known that oxygenated arterial blood is bright red whereas venous blood, which has lower oxygen levels in it, looks darker red and nearer blue. It is this colouration that makes veins near the surface look green or purple blue. Veins themselves are white when empty. When full of the darker red/blue blood, they tend to look green or blue/purple through a thin layer of skin.

Once the venous blood gets to the venous end of the capillaries, it is collected into small vessels called venules. Lots of venules then collect together becoming small veins which collect together into the larger veins. These larger veins drain into the large named veins, which then drain out of the leg and into the iliac veins in the pelvis, and from there into the inferior vena cava which transmits the venous blood back into the heart (right atrium).

The basic structure of the vein wall is the same as the arterial structure but has some modifications related to its function. It is much thinner as it does need not need to withstand the high pressures generated by the heart on the arterial side (**Figure 3**). Therefore the muscle layer, the middle of the three layers, is made up of fewer layers of smooth muscle than the arteries.

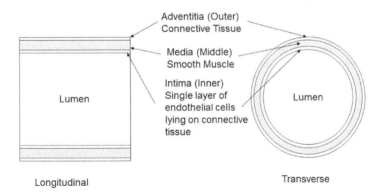

Figure 3: Basic structure of a vein wall. There are 3 layers, from outside to in – Adventitia, Media and Intima

Different veins do have different thicknesses of media (muscle layer) and this will be discussed elsewhere. In addition, the veins in the leg (and also in the arms) contain one-way valves, which the arteries do not. It is why these valves are required, as well as whether they are functioning or not, that is the whole basis of venous reflux disease. Therefore much of this book will be spent discussing these valves and their function - and so we will not delve any deeper at this point.

For completeness it is worth knowing that deoxygenated venous blood goes through the right ventricle and then, in systole, is forced into the pulmonary artery through the lungs where it gets rid of its carbon dioxide and takes on new oxygen, flowing back to the left side of the heart (the left atrium) via the pulmonary vein.

When learning about the circulation there are several interesting points to note:

- The pulmonary artery is the only artery in the body that carries deoxygenated blood, which would usually be called 'venous blood'

- The pulmonary vein is the only vein in the body that carries oxygenated blood, which would usually be called 'arterial blood'

- The left side of the heart has the same volume as the right side of the heart as they have to pump the same volume per minute in the same number of beats

- The left side of the heart has much thicker muscle as it has to pump blood at a much higher pressure to get it to circulate throughout the whole of the body, compared to the right side of the heart which only needs to push blood through the lungs

- Although carbon dioxide and oxygen exchange in the lungs (the pulmonary circulation), waste products are extracted from the blood in both the liver and kidney (systemic circulation) and nutrients are picked up by the blood in the gastrointestinal system (portal circulation)

Although this is the standard description of circulation, to understand the venous system of the legs and how blood returns, we need to start thinking about pressure and how it varies around the circulation as well as the effects of gravity and position of the person.

Let's start off by considering the pressure changes in the blood as it goes round the systemic circulation.

# CHAPTER 2

# Pressure in the venous circulation at rest

Having understood the basics of the systemic circulation, we can now move on to the more thorny issue of venous circulation and venous return from the lower leg. The easiest way to understand this is to take it in a stepwise progression.

To begin, we will consider the supine person and think of all of the pressure changes as the blood flows from heart, through the arteries and arterioles, then through the capillaries and finally back through the veins to the heart again.

Firstly, we need to have an understanding of what pressure is in the circulation as related to blood flow. Pressure is supplied to the blood by the heart muscle (myocardium) so that it flows forward. However, to understand the venous circulation fully, we must have a deeper understanding of the physics of pressure.

### 'Pressure is potential energy'

When the pressure is measured in any part of the blood at any time, it is a measure of the potential energy that the blood has at that point.

As the blood is ejected from the left ventricle through the aortic valve, it has the highest pressure, usually 120mmHg **(Figure 4)**. At this point the blood is full of energy, which it will need to force its way around the circulation and come back to the right atrium. By the time it returns to the right atrium, it will have used all of the energy up and therefore the pressure will be 0 mmHg.

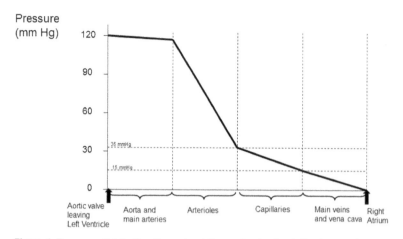

Figure 4: Pressure distribution throughout the systemic circulation in a supine person (ie: when lying down)

As we know from Sir Isaac Newton, energy cannot be created nor destroyed. It can only be converted from one form to another. Therefore, the pressure measured in mmHg in the arterial blood is a store of potential energy, which is used to move the blood through the circulation against the opposing factors. As it moves through these opposing factors the potential energy is used up overcoming these factors and is transferred to kinetic energy. This kinetic energy is then lost from the blood as vibration and heat.

The opposing factors to the flow of blood in the supine patient are:

• Inertia

• Resistance

Inertia is the term given to the resistance by an object to start moving. In other words when the blood is stationary in the heart or aorta, it takes energy to start it moving. Fortunately, compared to the other factors, inertia causes only minor energy loss in the circulation.

Resistance is mainly due to the friction of the blood flowing past the blood vessel walls. This will become clearer if we follow the loss of potential energy i.e. the pressure drop of the blood as it passes from left heart to right heart in **Figure 4**.

Blood is ejected from the aortic valve with a maximum systolic pressure of 120 mmHg and passes down the major arteries which have a very large diameter. As their diameter is large, the proportion of the blood flowing through the large lumen that is in contact with the wall is relatively small. Therefore there is a low resistance to the blood flow. As such there is only a small pressure drop as a large volume of blood flows through these large diameter vessels.

## Blood pressure readings from the arm

It is an interesting aside to note at this point in our exploration of the pressure changes through the systemic circulation that when we measure blood pressure in the arm, either using a sphygmomanometer or an invasive blood pressure monitoring system, we like to think that this pressure is the systolic pressure at the heart. Of course, it is not.

Blood would not have flowed to the arm if it had been the same pressure as when it left the heart. Blood, like all fluids, flows from an area of high pressure to an area of low pressure.

Therefore, the pressure we measure at the arm is lower than the pressure of the blood at the aortic valve as it leaves the heart. Fortunately for us, the arteries between aortic valve and the major arteries in the arm are all large and have low resistance. This means that minimal work is done by the blood in getting to our sphygmomanometer or pressure line, and so the potential energy (i.e. the measured blood pressure) is only a small amount lower than the pressure at the aortic valve. Thus, our measured pressure in the arm, although lower than the true systolic blood pressure at the aortic valve, is a good approximation.

## The Arterioles

As the blood flows on from the main arteries into the arterioles, there is a substantial change in this situation as can be seen by the dramatic pressure drop in **Figure 4**.

## Why is there this big pressure drop across the arterioles?

Clearly the arterioles must be high resistance, as the blood has to do a lot of work to get through them (this being reflected in the pressure drop). This pressure drop means the potential energy of the blood is being converted into kinetic energy as the blood forces its way through these vessels.

The arterioles keep the resistance high in the systemic circulation through two main factors:

- They are very narrow vessels

- There are relatively few arterioles

By keeping the diameter of each arteriole small, each arteriole has a large surface area compared to the total volume of blood flowing through it, causing friction with this flowing blood and therefore providing resistance to the flow. In addition, as the total number of arterioles is low compared to the large amount of blood coming out of the main arteries, the arterioles act as a 'bottleneck' through which the arterial blood has to squeeze, using potential energy and therefore dropping pressure in the process.

## What benefit do we get from having arterioles that cause a high resistance to blood flow?

Rather surprisingly, this high resistance network of arterioles is actually very useful for the body. It is by keeping all of the arterioles in the body at high resistance that the body can direct blood to where it is most needed.

To understand this consider eating. If you have just eaten a big meal, the gut needs an increased blood flow to feed the tissues which are working in the gut wall and to transport the nutrients digested from the food away for metabolism elsewhere in the body.

In order to achieve this, the parasympathetic nervous system and hormone system in the gastrointestinal tract act on the arterioles there, making them dilate. This dilatation reduces the resistance of the arterioles in the gut and so blood preferentially flows to the area of least resistance - increasing the blood supply to the gut and reducing it to the limbs and elsewhere.

Alternatively, if you decide to run, the muscles in your legs will need oxygen and nutrients. You start moving your legs fast and the carbon dioxide and metabolites in the working leg muscles start rising very quickly. These factors work directly on the arterioles in these muscles, causing them to dilate. Once again this dilatation reduces the resistance of the arterioles in the working muscles, causing blood to preferentially flow to them and away from other structures which aren't being used which still have higher resistance arterioles.

Of course in this second example of starting to run, there are other effects elsewhere in the body that help the blood flow more efficiently to the leg muscles. For example the adrenal glands start secreting the hormone adrenaline and the sympathetic nervous system becomes very active, increasing the 'sympathetic activity' such as in the 'fight and flight' reflex.

These factors lead to a slightly more complex situation where the heart beats quicker and with stronger contractions increasing the cardiac outflow by up to 6 times normal (from 5 L per minute in a normal man to 30 L per minute maximum), whilst also increasing the blood pressure transiently (to increase the potential energy to help the blood flow faster) and also constricting all of the other arterioles (i.e. to the gut or pelvic organs) maximising the blood flow to the required muscles.

So although this big drop of pressure across the arterioles looks problematic at first sight, it controls how the body ensures that blood flow ends up where it is needed at any particular time.

## Blood flow across the capillaries

In **Figure 4** you can see that arterial blood gets to the capillaries in the systemic circulation at a fairly constant pressure of about 35mmHg. It can also be seen from **Figure 4** that, as the blood flows across the capillaries, it drops its pressure from 35 mmHg to 15 mmHg due to the work it has to do forcing blood cells through tiny vessels.

These pressures should be remembered because they are essential when considering the microcirculation.

Looking at **Figure 5**, the hydrostatic (or pressure) forces on the blood in the capillaries can be seen. The walls of the capillaries are only one cell thick and therefore act as a 'micro filter' allowing water and very small ions and molecules across and into the tissue. Arterial blood coming into the arterial end of the capillary does so at a pressure of 35 mmHg. The fluid part of the blood gets forced out of the capillary lumen through the capillary wall, into the surrounding tissue fluid and from there diffusing into the local cells. This is how the nutrients are delivered from the blood to the tissues and cells.

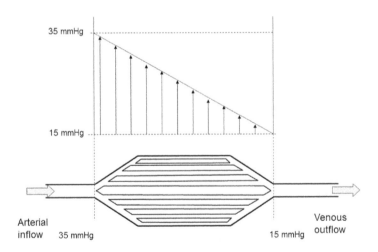

Figure 5: Hydrostatic pressures across the capillaries in legs in a supine person

The red blood cells are too large to get across the capillary walls and so they get caught in the lumen and stay within the capillary. As it is the red blood cells that carry the haemoglobin, which in turn carries the oxygen, the oxygen diffuses from the haemoglobin passing through the wall of the capillaries and into the local tissue fluid and cells as well.

As the blood flows across the capillary, resistance to the passage of the blood and, particularly, the red blood cells, causes a drop in the pressure resulting in only 15 mmHg hydrostatic pressure at the venous end. As can be seen in **Figure 5**, the drop from 35 mmHg at the arterial end of the capillary to 15 mmHg at the venous end causes a pressure gradient from capillary lumen to tissue fluid and so we could expect fluid to continue to leave the capillary blood all the way along the capillary.

However, if this was the only force affecting the capillaries the oxygen and nutrients would be delivered to the tissues and cells, but the carbon dioxide and waste products that the tissues and cells have produced by metabolism would not be removed.

The additional factor that corrects this and makes the removal of these waste products possible is the osmotic pressure of the plasma. This major pressure opposing the outflow of water from the capillary is known medically as the 'Plasma Oncotic Pressure', as shown in **Figure 6**.

Osmosis is a process whereby water moves passively (i.e. it is not pumped) from an area of low concentration to an area of high concentration through a semi-permeable membrane – in this case the capillary wall.

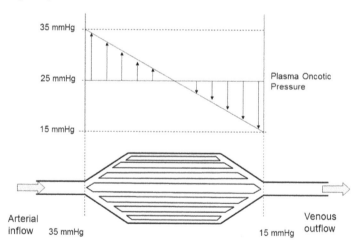

Figure 6: Resultant pressures across the capillaries in legs when plasma oncotic pressure is added to hydrostatic pressure (see text)

The plasma (the liquid part of blood) is full of proteins that cannot get through the capillary wall and are therefore stuck inside the capillary. The proteins mixed with the water in the plasma are therefore an area of high concentration which can have an 'osmotic effect', i.e. attract water into it through the capillary wall. To know whether water will pass into or out of this solution, we need to know what level this osmotic effect is, or in other words the plasma oncotic pressure. This can be measured in mmHg. The plasma oncotic pressure in the capillaries is 25 mmHg, halfway between the arterial and venous hydrostatic pressures.

The result of these opposing pressures – the hydrostatic pressure forcing water out of the capillary and the plasma oncotic pressure attracting it back inwards - can be seen in **Figure 6**. In the first half of the capillary, fluid moves out of the capillary into the tissues by hydrostatic pressure as described above. At the halfway mark, the tipping point occurs when there is no net flow across the capillary wall. After this point, the plasma oncotic pressure is higher than the hydrostatic pressure and water moves back in from the tissue into the capillary once again. As it moves back into the capillary, it brings with it carbon dioxide and the waste products of metabolism from the surrounding cells.

These pressures and the resulting flows of water, dissolved gases, nutrients and metabolites are essential to the healthy functioning of tissues. If any of these change, there will be an interruption of the very fine balance in providing oxygen and nutrients to the living cells and the removal of carbon dioxide and waste products.

Low blood pressure or low-protein states can have a dramatic effect on this balance but are not relevant for this book. However, an increased pressure at the venous end of the capillary, or an inflammation in the wall of the venous end of the capillary preventing easy flow of fluid, have significant effects. As these are found in venous reflux disease we will return to these later in the book.

### Venous return from venules to the right atrium

#### In the supine position (Figure 7).

As discussed above, once the blood has managed to get through the capillaries it enters the venules. At this point, the pressure of the blood is approximately 15 mmHg.

If the person is lying down, the pressure in the right atrium is usually 0 mmHg, although this does change a little with contraction of the atrium and also with breathing. However, to understand the physiology of venous return from the lower limb it is adequate to take the pressure to be relatively constant at 0 mmHg.

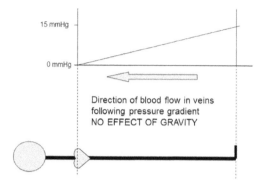

15 mmHg

0 mmHg

Direction of blood flow in veins
following pressure gradient
NO EFFECT OF GRAVITY

Figure 7: Venous return from feet to heart in a supine person

Therefore, considering blood in the venules at the foot or ankle, there is a 15 mmHg pressure gradient driving blood through the veins towards the heart. As the veins are a very low resistance network in the normal person, that readily dilate if extra capacity is required, this is more than sufficient energy to ensure that blood flows from feet to heart without problem.

### Breathing and venous blood flow in the supine position.

As anyone who has performed any venous studies of the flow of blood in a normal person has observed, the flow of venous blood in the veins of the legs isn't constant, even when the patient is lying still. There is a change in the flow of blood in the leg veins that is coordinated with respiration.

When the patient is breathing out (expiration), pressure in the chest increases, forcing air out of the lungs. This increased pressure within the chest reduces the pressure gradient from the ankle to heart. The pressure in the venules at the ankle remains at 15 mmHg but the pressure within the chest is now higher, therefore slowing or even stopping the flow of blood from foot to heart.

When the patient is breathing in (inspiration), the chest expands and a negative pressure is generated within the chest cavity in order for air to flow into the lungs from the atmosphere. Within the venous system of the leg, this increases the gradient of pressure from venules to heart, causing an increase in the blood flow in the veins of the leg, pelvis and abdomen.

### In the standing position (Figure 8)

Up until this point the whole of the circulatory system has been quite simple to understand. It is when a person stands up that even

18

the most experienced healthcare professional's understanding starts wavering; especially when it comes to understanding what happens when parts of the system go wrong, as in venous reflux disease (chronic venous insufficiency). Therefore, although the arguments leading up to this point may have appeared basic, it is well worth reading both them and the following, in order and in detail, to make sure that you get a full grasp of all of the concepts relating to this common problem.

On standing, there is in effect a column of blood extending from ankle to heart. (Please remember that all pressure measurements are measured with relation to the heart, the haemodynamic centre of the patient, not the head. Therefore when we talk about height later on, we are not talking about the height of the patient but the height of the heart above the feet. The pressure in the head is lower than the pressure in the heart when standing due to the same gravitational effects that we will discuss below, but in reverse. This does cause many interesting effects for neurologists and physicians but is not a subject we would consider any further in our study of phlebology). This column of blood exists on both sides of the circulation, arterial and venous.

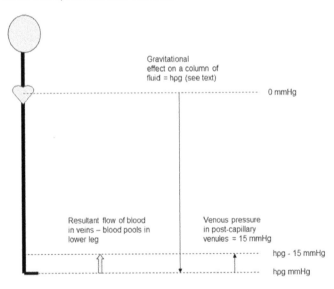

Figure 8: Pressure in the venous blood in an erect person (i.e. standing still)

On the arterial side the increased pressure in the column of blood, at any point of the column, aids the forward flow of the arterial blood away from the heart. At the capillary level, as the tissue fluid pressure is raised at the same level as the pressure of blood within the capillaries, both due to gravity, there is no change in the overall function of the capillaries due to this pressure alone. It is on the venous side that things become interesting.

The pressure of the column of blood from foot to heart depends on the height of the patient - more specifically the height of the patient's heart above their feet. The exact pressure at the bottom of the column (i.e. at the ankle/foot) can be calculated by the simple formula:

Pressure = hpg

**Where**

h = the height of the heart above the feet

p = the density of the blood (remembering that density is basically the 'weight', or more correctly 'mass' of a certain volume of blood)

g = gravity

Therefore this column of blood produces a significant pressure on the venous blood in the venule.

Opposing this is the potential energy, or pressure, of the blood emerging from the capillaries into the venules. As we have outlined previously, this equates to approximately 15 mmHg. This pressure is sufficient to drive the blood from the foot and ankle to the lower leg, usually just below the calf. At this point the potential energy is spent driving the blood against gravity and the venous blood cannot flow any further by this hydrostatic pressure alone.

It is for this reason that soldiers standing without any movement whatsoever on the parade ground often faint; as would anyone who simply stands completely still for any length of time. If you watch people standing for any period of time, you will notice that they are always making small movements, shifting their weight from one leg to the other. Soldiers are taught to wiggle their toes in their boots, or gently shift their weight from one foot to the other, as they are standing to prevent them from fainting. The same can be seen in people standing up at a sink or when doing a chore such as ironing.

It is this movement that pumps the venous blood back from where it is trapped in the lower leg to the heart against the force of gravity when we are standing.

The venous pump is all important to humans to allow us to stand on our feet for long periods of time without fainting, suffering from swollen ankles, or any of the other problems that we will be exploring later in this book. It is the basis of phlebology and understanding what happens when the venous pump fails.

# CHAPTER 3

# Functional anatomy relating to the venous system of the legs

To understand how the venous system works, and more importantly what happens when parts of it fail such as in venous reflux disease (chronic venous insufficiency), it is essential to have a concept of the anatomy of the veins not only in the legs, but those relating to the function of the legs.

This book sets out to elucidate venous function in the normal leg and promote the understanding of problems when this venous function fails. Therefore, rather than go into the rather complex anatomy and anatomical variations which may occur in life (these will be covered in other books in this series), we are going to concentrate on the arguments related to basic venous anatomy.

This means that we will concentrate on the main veins that are important in venous reflux disease. The principles and arguments that we explore will describe the function and effects of the failure of the function of these veins. Once these are fully understood, the principles can easily be transferred to specific veins that are the problem in individual patients.

We will consider the anatomy of the vein itself and the venous valves, followed by the anatomy of how these veins are then arranged in the body.

## Structure of veins and venous valves

### Vein wall

Chapter 1 outlined the basic structure of the venous wall (**Figure 3**). All blood vessels in the human body have the same basic structure, they just vary in size and relative proportions of each of the different components.

## Basic structure of blood vessels – arteries and veins:

Veins are thin-walled blood vessels compared to arteries. However they have the three same basic layers:

- Adventitia - on the outside

- Media - the middle of the three layers of the wall which contains smooth-muscle

- Intima - the inner layer which has a single layer of endothelial cells on its inner surface

The endothelial cells on the inside of the intima have the special property of being very flat, which allows the blood to pass over them. A healthy, intact layer of endothelial cells will not allow any blood passing over them to clot (which would be called a 'thrombosis'). Clots or thromboses in the veins are often caused by damage to the endothelial layer. When a clot forms in a vein it causes a deep vein thrombosis (DVT), or a superficial thrombophlebitis (Phlebitis) depending upon which vein is affected.

It should be noted here that different veins have vastly different sizes: ranging from the venules at the venous end of the capillary network, which might be less than half a millimetre diameter; through to the truncal veins of the legs such as the Great Saphenous Vein, which might be about 5 mm in diameter (when normal); or the inferior vena cava which ranges between 10 to 18 mm diameter depending on the size of patient. All of these different sizes of veins have different proportions of smooth muscle cells in their walls and therefore different thicknesses in vein walls predominantly related to the thickness of the media.

At this point, we do not need to explore this further. However, it does become vitally important when we are considering different treatment options and want to ensure that the treatment we are using will penetrate the whole of the vein wall, destroying all of the cells from endothelium to the adventitia.

## Venous valves (Figure 9)

It is the presence of valves on the inside of the vein wall that makes the veins different to the arteries at a basic level. As we shall discuss in the next chapter, the valves are basically an anti-gravity device which prevents the blood from falling down the veins and away from the heart on standing.

The basic structure of each venous valve is quite simple - although in common with a lot of medicine if studied carefully and in great detail, there is a huge complexity in its structure that allows its function to work so well. However for the purposes of this book, we only need to have an understanding of the basic structure and function of these valves.

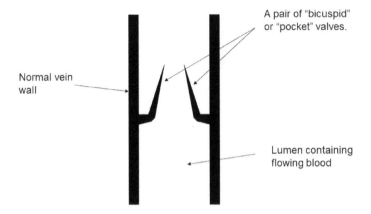

Figure 9: Normal valves in leg veins

Firstly, the valves are bicuspid. This means that they are arranged in pairs on opposite sides of the vein wall. They are like little pockets of tissue protruding from the wall into the lumen of the vein, with the open part of the pocket pointing upwards and the closed part pointing downwards.

The valve leaflets are very thin, coated on each side with endothelial cells to stop any clotting and have a very thin core of intima to provide support and nutrition for the endothelial layer. As the vein wall itself can expand and contract these valves have to be very pliable to continue working as the vein changes its own diameter. They need to keep working even when they may be compressed by the action of muscles on the outside of the vein wall.

Considering a venous valve somewhere in the mid leg, when blood is being forced upwards through the valve, the valve leaflets are pressed against the wall, allowing a smooth blood flow upwards with almost no turbulence at all (**Figure 10**).

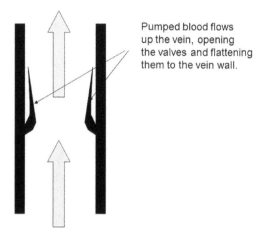

Pumped blood flows up the vein, opening the valves and flattening them to the vein wall.

Figure 10: Normal valves in leg veins – opening during active muscle contraction lower down the leg forcing blood up the vein

However when blood falls down the same vein due to the effect of gravity, it hits the leading edge of the bicuspid valves, opening the valve and getting caught in the blind ending pocket (**Figure 11**). As both sides of the valve open simultaneously, they meet in the middle, preventing any flow whatsoever down the vein at this point.

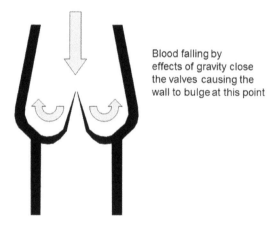

Blood falling by effects of gravity close the valves causing the wall to bulge at this point

Figure 11: Normal valves in leg veins – closing during passive phase when blood refluxes or 'falls' back down veins by effects of gravity

This appears very simple but it is wise to note at this point that:

- The opening and closing of the valves is a passive action of the valve responding to the flow of blood through the vein, and importantly:

- There has to be some flow backwards down the vein to close the valve before the backward flow is completely stopped.

One final thing to note is that although all normal veins in the legs have venous valves, it is rare to find any that work in the pelvic (iliac) veins or the inferior vena cava. This becomes very important later when considering the mechanism of venous reflux disease (chronic venous insufficiency).

### Functional anatomy of the venous system

### A note on names of veins

Before we proceed it is important to note that up until 2001 many different names were used for the same veins. For example, the main superficial venous trunk that starts by draining veins from the foot of the ankle on the medial (inner) side, and ascends on the inner aspect of the leg usually just behind the knee, before joining the deep vein at the groin has been known as the:

Long saphenous vein
Greater saphenous vein
Great Saphenous Vein
Saphena magna

as well as several other less well-known names.

Similarly the truncal vein arising by draining veins from the foot at the ankle on the lateral side, and then ascending in the superficial tissues at the back of the leg in the midline, inserting into the deep vein behind the knee has been known as the:

Short saphenous vein
Lesser saphenous vein
Small Saphenous Vein
Saphena minima

as well as other less well-known names.

In the academic world of phlebology it had been a constant irritation that different textbooks, research papers or presentations use different names for the same structure. In particular, this became very confusing when people used abbreviations. The simplest example is when people were referring to the long saphenous vein as the LSV, whilst other people were referring to the lesser saphenous vein, a completely different vein, but also using LSV.

In 2001 at a European meeting in Rome, it was decided to standardise all the names of the veins. This nomenclature was accepted by American phlebologists in 2004 and has been used in academic circles and in good vein units ever since.

This book will only use these new names. It is a constant surprise to me how many research papers that I am asked to review, and the many presentations I sit through at venous meetings, in which the old names are still used by people purporting to be venous researchers or experts.

Hopefully as education and interest in phlebology becomes more widespread, the use of the correct names will also increase.

## The layout of veins in the leg (Figure 12)

Classically the veins in the leg have been divided into

• Deep veins

• Superficial veins

Some of the worst misunderstandings about venous disease over the last hundred years have come from a belief that these are two completely separate systems. Indeed it used to be thought that failure of the valves in the deep system of veins caused leg ulcers whilst failure of valves in the superficial system cause cosmetic varicose veins only. This overly simplistic view was very popular in the 1960s and 1970s but was shown to be completely wrong in the mid-1980s.

One of the first concepts that is important to consider is that the main veins, both superficial and deep, are draining venous blood from the veins of the foot. At the ankle, most of the diameter of the leg is bone. At this point and in the foot itself, the deep and superficial systems are connected, meaning that the old concept that one could have high-pressure in the deep system without it affecting the pressure in the superficial system, or vice versa, are clearly incorrect.

With a good knowledge of haemodynamics this should have been obvious, but many research studies have now confirmed this, to convince even the sceptics who still like to try to treat patients as if the deep and superficial systems are completely separate.

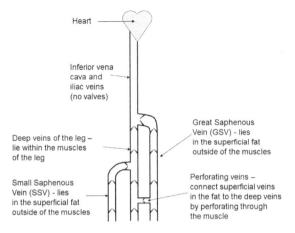

Figure 12: Normal functional anatomy of veins draining blood from the leg

## The deep veins

For the understanding of this book, we will consider all of the deep veins as one big deep venous trunk. There will be further books in this series that go into the anatomy in more detail but for those who want to have some deeper anatomy at this point, there are three pairs of veins in the lower leg (the anterior and posterior tibial veins and the peroneal veins), all surrounded by muscles in the lower leg, that all join together below the knee forming the popliteal vein.

This vein runs from behind the knee joint around the bone in the thigh on the inner aspect and deep within the muscles of the thigh, leaving the leg just above the groin. Traditionally it is called the popliteal vein until it winds around the thigh bone (femur) when it then becomes called the femoral vein. Many phlebologists call the whole segment from the junction of the three pairs of lower leg veins to the groin the femoropopliteal vein.

As the femoropopliteal vein leaves the leg in the groin, it passes under the inguinal ligament and at this point is called the iliac vein which runs towards the midline and then meets the same vein from the other side at a level just below the umbilicus (belly button) to become the inferior vena cava. This then runs up through the diaphragm, into the chest and into the heart at the right atrium.

The deep veins are surrounded in muscle from ankle to groin. This becomes important in the understanding of the venous pump, which will be discussed in the next chapter.

All of the veins in the deep system contain bicuspid venous valves. They tend to be approximately 8 cm apart although this can vary widely.

## The superficial veins

The superficial veins are called superficial not because they are visible on the surface, but because they lie outside of the muscle of the leg in the subcutaneous fat. There are some complex relationships with connective tissue called 'fascia' in this fat that are important surgically and these will be discussed in further books in the series.

## The truncal veins – Great Saphenous Vein (GSV) and Small Saphenous Veins (SSV)

For the understanding required of the venous pump and venous reflux disease (chronic venous insufficiency) we can concentrate on the two main superficial veins in the leg.

As a point of interest, embryologically both the arm and leg have two main superficial veins, one on each side of the limb bud. These become the two main superficial veins or 'venous trunks' once born, becoming the Great Saphenous Vein (GSV) and the Small Saphenous Vein (SSV). In the lower limb bud, as the leg develops, the limb bud rotates inwards bringing the great toe which is the top part of the limb bud to being front and medial (towards the inner part).

The vein that was on the top of the limb bud becomes the GSV and now runs from the front of the bony part of the inner ankle (in front of the medial malleolus) up the inner part of the lower leg, past the knee on the inner aspect and just behind the joint, and then up the thigh reaching the groin on the front of the leg. At this point, the vein dives deep through the muscle joining the femoropopliteal vein in the groin at a junction called the saphenofemoral junction (SFJ).

The other main superficial vein of the limb bud becomes the SSV and now runs from behind the bony part of the outer ankle (behind the lateral malleolus). It then ascends the lower leg in the midline until somewhere behind the knee, where it usually dives deep and joins the popliteal vein at a junction called the saphenopopliteal junction (SPJ).

It is very important to note at this point that, despite both of these major truncal veins being called 'superficial' veins, they are rarely visible on the surface as they are deep within the fat layers and surrounded by their own special envelope of connective tissue called fascia. This point is not understood widely and many doctors and nurses who see veins on the surface lying in approximately these positions mistakenly think they are seeing actual truncal veins. Unless the patient is incredibly skinny, this is highly unlikely.

All of the veins in the superficial system contain bicuspid venous valves. The distance between them varies widely but can be approximately 10 cm in the main part of the truncal veins.

## Perforating veins

As seen from **Figure 12** the superficial veins have to connect to the deep veins in order to drain blood from superficial tissues into the deep veins and then back up to the heart. As above, the two truncal veins have their own junctions, which perforate the muscle and connect directly into the deep venous system.

In addition to these two specifically named junctions or 'perforating veins', there are approximately 150 smaller perforating veins that connect superficial veins with the deep veins, running through the muscles in the leg. Most of these perforating veins, or perforators as they are called surgically, occur in the lower leg below the knee.

The function of these perforating veins is to take blood from the superficial veins into the deep veins from where they can be pumped up to the heart by the venous pump. They have valves within them, making sure that it is a one-way stream of blood from superficial to deep.

## Inwards and upwards (Figure 13)

In the normal person's anatomy, as described above, the function of the venous system of the leg makes sure that the blood flows in a one-way system. All blood is collected and directed into the deep system and up to the heart. The mechanism of how the blood is pumped is the subject of the next chapter.

The consequences when this fails will be discussed later in this book at length.

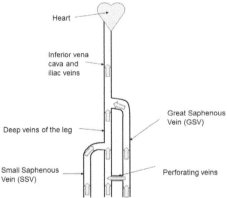

Figure 13: Normal direction of blood flow in the veins draining the leg during active pumping of venous blood – generally the flow is 'inwards and upwards'

# CHAPTER 4

# The venous pump of the lower limb

**The Venous Pump**

As we discussed in Chapter 2, if there was no external driving force to push blood up inside the veins from the foot and ankle towards the heart, blood would get no further than the lower leg with the hydrostatic pressure of the blood in the venules alone. The residual pressure in the blood, once it has traversed the capillary network, is approximately 15 mmHg.

Therefore, we need to pump the blood up out of the leg against gravity.

**Key components of a pump**

There are two main components required in a pump in order to pump fluids:

- A pumping force to provide energy to move the fluid in the active phase of the pumping

- A valve system to stop the fluid returning to its original position when the pumping force relaxes in the relaxation phase of the pumping.

In its simplest form, a pump transmits force into a fluid - in this case the fluid being pumped is blood. This force is stored within the blood as pressure, once again increasing the potential energy of the blood allowing it to start doing some work.

Clearly, if there were no exit from the pump, the blood would just be pressurised and would not move. Therefore the key to the active phase of a pump is that there must be both a mechanism to put energy into the blood and an exit (or low resistance route) that the blood can flow along.

The second most important feature is that once the blood has moved against gravity along the low resistance route, it is prevented from falling back down again when it runs out of energy.

## Pumping blood

When a non-expert is asked how venous blood returns from the leg in a person who is standing, they usually reply, "it returns due to the muscle pump'. Some will then include the negative pressure in the chest due to breathing, as a secondary effect, but this belies the fact that they do not understand the complexities of the muscle and foot pumps of the leg. We will go through these now.

In the leg, the deep veins are encased in muscle. It is this muscle which provides some, but not all, of the active pumping force. When the muscle compresses the deep veins during the active part of the pumping, theoretically blood being pressurised by the muscle could flow either upwards towards the heart or downwards away from heart. For several reasons the blood only flows upwards towards the heart.

Firstly, these veins are being fed from smaller veins, which are in turn fed from the venules which collect blood from the capillaries. Some of these smaller veins are not involved in the pump and therefore are not being compressed by the muscles themselves. Neither do they contain valves. Although the lack of valves in these small veins does not stop blood under pressure flowing backwards into the venules, both the inertia of the flow of blood forwards, and the residual hydrodrostatic pressure of the blood in the venules have a small effect reducing the flow backwards towards the capillaries.

Secondly, blood that does flow back towards the capillaries meets a higher resistance bed than the alternative route up the increasingly wider veins towards the heart.

Thirdly, a pump called 'the foot pump' has already pumped blood reaching the deep veins in the muscle pumps in the leg. Hence there is a forward motion and inertia of the blood entering the deep veins from the pumping action of the foot pump.

The role of the foot pump has really only been understood over the last 10 to 20 years and has been one of the major changes in our understanding of how the venous system works.

## The foot pump

One of the reasons that it has taken so long to understand the foot pump is that it has been so easy to understand the muscle pumps in the leg that doctors have not looked further than these. People without any medical background are well aware that muscles contract and can do so forcibly. One can easily picture how the contraction of the muscle pressurises the blood within the vein. The addition of valves within the vein above, below and within the section being pressurised allows simple understanding of how this pressurisation of the blood is converted into a one-way pump.

The foot is much more difficult to understand at first sight, although I hope by the end of this section it will appear clear to you.

The foot has very little muscle within it and most of the veins lie outside of the muscle. It is made up of a series of bones which are held by tendons ligaments, as well as a tight band called the plantar fascia into a pair of arches.

The first arch extends from front to back, with the heel and ball of the foot (the calcanium and the heads of the metatarsals) in contact with the floor with the rest of the foot bones forming an arch like an archer's bow (**Figure 14**). There is also a second arch which is transverse across the foot, with the outside of the fifth metatarsal being in contact with the ground and the other metatarsals forming an arch away from the ground, the other end of this transverse arch being the ball of the foot.

Having a foot arch that is held in place by tendons, ligaments and the plantar fascia is very similar to having an archer's bow with the wooden part of the bow being equivalent to the bones, and the string of the bow being equivalent to the other structures. At rest, there is a potential energy in the arch but this cannot be released or used without an action.

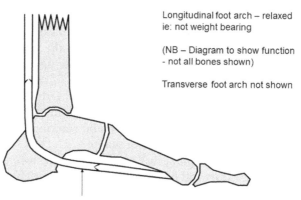

Longitudinal foot arch – relaxed
ie: not weight bearing

(NB – Diagram to show function
- not all bones shown)

Transverse foot arch not shown

Veins slung between the foot arch in direct communication
with superficial and deep veins of leg

Figure 14: The foot pump in the relaxed state

The key to the foot pump is that the veins in the foot are suspended within the arch itself and so the tension in their walls is determined by the status of the foot arch.

When the bodyweight is shifted onto the foot, the weight is transmitted as a force down the tibia (shinbone) and is transmitted through the ankle joint to the bones of the foot (**Figure 15**). This then flattens the arch, stretching the veins slung under the arch, thereby increasing the tension in the vein walls and narrowing their diameter. This pressurises the blood within them which pumps the blood out of the foot, into veins around the ankles, and from there into the deep and superficial venous systems of the leg.

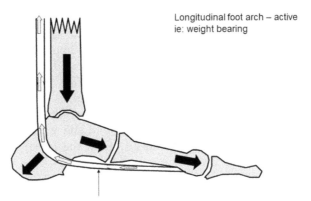

Longitudinal foot arch – active ie: weight bearing

Veins pulled and stretched, narrowing the lumen and pressurising the blood within the lumen, forcing blood out of the foot

Figure 15: The foot pump in the active (weight bearing) state

Once the weight is removed from the foot, the foot arch springs back to its normal shape, relaxing the veins and allowing them to fill with more venous blood from the post-capillary venules.

Although it is useful to think of an archer's bow to visualise the potential energy of an arch, the action of the foot pump is actually the opposite of the way a bow is used. When a bow is used to fire an arrow, a force is used to bend the string and thus increase the curvature of the bow. The release of the arrow enables the power of the bow returning to its less bent shape to be transmitted to the forward motion of the arrow.

The force of putting the bodyweight onto the foot is transmitted directly to the blood, causing the blood to pump when the arch of the foot (the bow) is somewhat flattened. When the bodyweight is released, the arch can spring back to its normal shape, allowing the veins to relax. This is where the comparison to the bow fails: the forward motion of the blood occurs when the external force is applied to the foot arch rather than on its release as with a bow and arrow.

The foot pump has a smaller volume than the muscle pumps as it is the furthest part of the leg away from the heart, and therefore has fewer veins and capillaries feeding blood into it than more proximal parts of the leg. And yet, research into the strength of the foot pump suggests that it is in fact the strongest, and some think the most important, of the venous pumps in the leg.

Not only is it a very important pump in terms of its strength and efficiency, but it is the most distal of the pumps and therefore is responsible for clearing the venous blood from the foot. In addition, it is the first of the venous pumps to be activated in the coordinated pumping action that we call 'walking'. This will be discussed below.

Finally, although Figures 14 and 15 show the arch between heel and metatarsals, it must not be forgotten how important the lateral arch is. This arch is between the ball of the foot (the first metatarsal head) and the 'lateral' ball of the foot (the fifth metatarsal head). When standing, the heel is on the ground and so both of the foot pumps are activated. When running on the toes (the correct way to run) the fore foot is placed on the ground and is flattened – the foot pump now being pre-dominantly from the lateral arch being stretched flat.

## The muscle pumps

Although there are several muscle groups containing veins in the leg, most authorities seem to favour the idea that there are three main muscle pumps in the leg, in addition to the foot pump described above.

The first two muscle pumps in terms of sequence of action are the deep and superficial calf pumps. As these two pumps act together whenever the calf is contracting, some people maintain this is one pump. However, in view of the two distinct muscle groups, gastrocnaemius and soleus, each with a venous network within them, it is more correct to consider them as two separate muscle pumps in the calf working in synergy.

The third and final muscle pump in the leg is the thigh muscle pump that is mainly in the quadriceps muscle on the front of the thigh.

Although it is customary to think of these muscle pumps individually and just accept that this is how blood returns to the heart, to do so is missing the beauty of the coordinated pumping action that is achieved by walking.

In addition, the co-ordinated way in which the pumps should work becomes very important later on when considering some of the problems that can occur with venous reflux disease (chronic venous insufficiency); particularly in the 'double whammy effect' that has not been previously described before this book.

## Walking – co-ordinated venous pumping from the leg:

All four of the vein pumps come together in a wonderfully co-ordinated mechanism when walking. We will go through these with the aid of the Figures 16 to 19.

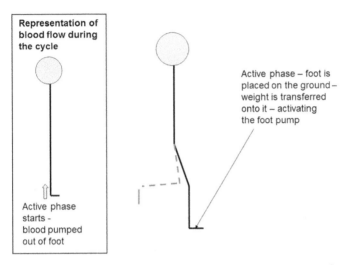

Representation of blood flow during the cycle

Active phase – foot is placed on the ground – weight is transferred onto it – activating the foot pump

Active phase starts - blood pumped out of foot

Figure 16: Walking – a co-ordinated venous pump - 1st pump = The foot pump

The first action that starts the vein pump is when the leading foot is placed on the ground (**Figure 16**). As the foot is placed on the ground, the bodyweight passes on to it, flattening the foot arch and activating the foot pump. Venous blood that has been collecting in the foot veins gets pumped at high pressure out of the foot, via the veins in the ankle and into the deep veins of the lower leg (and a small amount into the superficial veins at the ankle communication).

As the bodyweight continues to move forwards, the other leg is being brought forwards ready to take the next step and the leg that we are interested in is passing under the body (**Figure 17**). As the weight passes forwards, the heel lifts off the ground and the bodyweight is passed onto the toes. As this happens, the foot is actively extended to 'spring' the body forward so that the next step is longer than it would otherwise be. This extension of the foot is called the 'toe off' and is caused by contraction of the calf muscles.

This activates the calf pump, both superficial and deep, which continues pumping the blood that has come from the foot as well as adding more venous blood from the tissue of the lower leg itself.

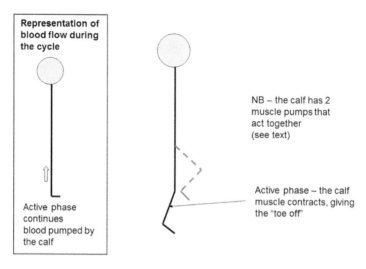

Figure 17: Walking – a co-ordinated venous pump - 2nd (and 3rd) pump(s) = Calf pump and the 'toe off'

As the other leg starts its cycle, with the foot being placed on the ground, the bodyweight is taken off the foot that we are looking at (**Figure 18**). The non-weight-bearing leg is lifted and brought forwards mainly by the thigh muscles, to position the leg and foot ready to take the next step. The contraction of the thigh muscles in this process is the fourth and upper most of the venous pumps, continuing the coordinated pumping of blood from foot to pelvis.

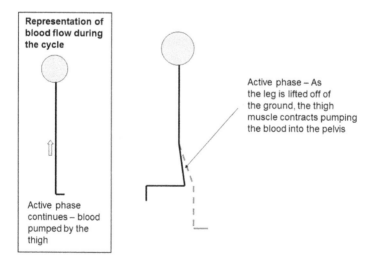

Figure 18: Walking – a co-ordinated venous pump - 4th pump = The thigh pump

Finally, there is a passive phase when the leg is swinging into position without much muscle activity (**Figure 19**).

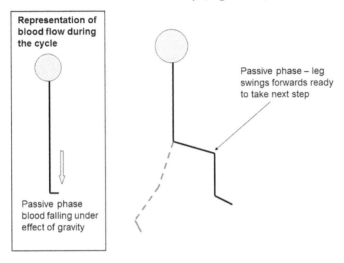

Figure 19: Walking – a co-ordinated venous pump – The passive phase

At this point blood that has reached the pelvis already has some inertia to keep taking it towards the heart. In other words, it has potential energy supplied from the coordinated muscle pumps that can be measured as pressure. This is aided by inspiration which causes negative pressure in the chest and therefore increases the pressure gradient in favour of blood continuing to flow towards the heart.

In expiration this gradient is worsened and so blood will only continue to flow towards the heart during that part of the breathing cycle if the potential energy supplied to the blood by the muscle pumps (i.e. the pressure of the blood entering the pelvic veins), is higher than the gravitational pressure of the column of blood above it and the extra pressure created in the chest by expiration.

Should it be that during the passive phase of the walking cycle the pressure of the blood entering the pelvic veins is not sufficient to overcome the gravitational pressure and the pressure within the chest caused by expiration, the blood will not continue to flow towards the heart; it will reverse and try to fall back down into the veins in the leg. This backward flow of blood is called venous 'reflux'.

As has been noted before, there are usually no working valves in the inferior vena cava or the iliac veins in the pelvis. Therefore, there is nothing to stop this blood falling back downwards or 'refluxing' down to the femoral vein at the groin.

Fortunately in the normal person, the top valve at the saphenofemoral junction stops blood from refluxing into the Great Saphenous Vein, and the top valve in the femoral vein stops blood refluxing down the deep veins. This is why, for over 100 years, doctors have thought that varicose veins are caused by the failure of the top valve in the Great Saphenous Vein. This is incorrect, as my research group published in 2001, and the actual process will become clear in the next few chapters.

## Important factors to consider when thinking about the venous pump

Although the understanding of the foot pump and therefore the co-ordination of the venous pumps have only been fully understood over the last 10 to 20 years, there are of course further complexities that may or may not have an effect on the venous pump. The above description has concentrated on one leg during the walking cycle; we need to consider the second leg. In the passive phase of the walking cycle when blood may reflux back down the iliac vein on the side we are thinking of, there is active pumping of blood in the iliac vein on the other side at the same time which forces blood at high pressure up towards the heart.

This means that it is likely that blood in the inferior vena cava is still moving towards the heart despite blood potentially refluxing in the iliac vein on the relaxed side. The speed of blood passing up the active side will pass across the open end of the iliac vein on the relaxed side, reducing the pressure and having somewhat of a 'siphon effect'.

These different effects have not been fully elucidated and, apart from an academic interest to understand fully how these things work in the normal person, they have little effect on the subsequent arguments as to when things start going wrong with the venous system resulting in venous reflux disease (chronic venous insufficiency).

Therefore if you have understood the four different pumps, how they are coordinated in walking and, most importantly of all, have understood that there is an active phase followed by a passive phase in the walking cycle, then you are ready to move on to the next chapters where we will start to explore what goes wrong and why different people with the same underlying problem can have completely different clinical problems.

# CHAPTER 5

# Valve failure and venous reflux (venous insufficiency)

At the end of the 19th century, a German doctor named Trendelenburg performed post-mortem examinations to find out the causes of varicose veins. He studied people who had normal leg veins in life and on those who had suffered from varicose veins in life to identify what the underlying difference was between them and to identify the cause of varicose veins.

He found that those people who had normal leg veins had the normal bicuspid venous valves in their deep and superficial veins. However in those people who had suffered with varicose veins when alive, he found that the normal bicuspid venous valves had failed in the superficial veins and were not able to function (**Figure 20**).

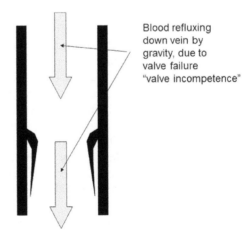

Blood refluxing down vein by gravity, due to valve failure "valve incompetence"

Figure 20: Incompetent valve in a leg vein – allowing venous blood to reflux during passive phase due to gravity

Having made this observation, he then proceeded to make educated assumptions or guesses as to the function of these valves in life and what happened when they failed. This led him to invent a clinical test for varicose veins called the 'Trendelenburg test', as well as to invent an operation for varicose veins called the 'Trendelenburg operation' which we would now call the 'high saphenous tie', or 'crossectomy'.

Unfortunately, although his observations were accurate in the dead people he was examining, his assumptions as to how his findings translated to the problem of varicose veins in the living were erroneous. As no-one else had advanced any other theory, his assumptions became widely accepted as the basis of varicose vein formation. This led to a general misunderstanding about the development of varicose veins and thus led to poor treatments over the next 100 years.

Despite being wrong in his assumptions as to function, nothing should be taken away from Trendelenburg as he did not have the techniques, such as duplex ultrasound, that we have nowadays to examine how venous valves function in the living patient. However, the same cannot be said for those who perpetuate the myths and erroneous teachings that have sprung from Trendelenburg's observations.

We will now go through these misunderstandings that are still taught by many, in a stepwise fashion, so that you can understand the mistake in Trendelenburg's assumption and start to see the consequences this has had on traditional varicose vein surgery.

### The 'domino theory' of venous reflux

If, like Trendelenburg, we only had post-mortem specimens to work with, we would only know the anatomy of the veins and whether the valves appeared to be intact or not. We would be able to show that in people who had no venous disease in life that the valves were normal at post mortem. However in people with varicose veins in life (one of the clinical manifestations of venous reflux disease or chronic venous incompetence) the valves would be found to be faulty and unable to work. This being the case we would probably all come up with the domino theory as the most likely cause of the development of the venous reflux – which explains why this has been believed by most doctors and widely taught over the last century.

To understand the domino theory, please refer to the three sequential diagrams in **Figure 21**. In addition, remember that the domino theory and the understanding of venous reflux by the majority of doctors and researchers, only considers the passive phase

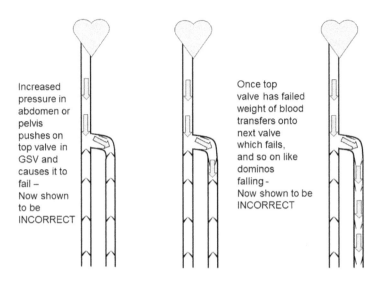

Increased pressure in abdomen or pelvis pushes on top valve in GSV and causes it to fail – Now shown to be INCORRECT

Once top valve has failed weight of blood transfers onto next valve which fails, and so on like dominos falling - Now shown to be INCORRECT

Figure 21: 'Domino theory' of development of venous reflux and varicose veins – now proven to be INCORRECT

of the venous pump. This is, therefore, represented by someone who has been lying down and who suddenly stands up and does not move or activate any of the vein pumps.

The concept of the domino theory is that the column of blood from the groin to the heart puts a gravitational pressure on the top valves of both the deep system and the Great Saphenous Vein at the groin. It supposes that the valve at the top of the Great Saphenous Vein, the valve of the saphenofemoral junction, eventually gives way and fails. This allows the column of blood to exert its pressure on the next valve down the Great Saphenous Vein. Once again this is thought to fail and so on, until all the valves of the Great Saphenous Vein have become incompetent. Thus the blood can reflux freely down the whole length of the vein on standing.

This is an obvious cause of the reflux that satisfies all of the early observations - but unfortunately subsequent research has shown it to be completely wrong, as will be explained later in this chapter.

### Increased abdominal and pelvic pressure

One of the consequences of the domino theory of venous reflux is that doctors have thought up various conditions that increase the pressure within the abdomen and pelvis; which (to their understanding) would increase the pressure of this column of blood over and above its weight alone and therefore would hasten the failure of the valves. This in turn would hasten the development and progression of the venous incompetence down the whole Great Saphenous Vein.

Thus in the older and out-of-date textbooks and articles, it is common to read that pelvic tumours, obesity and even constipation cause varicose veins by an increase in abdominal and pelvic pressure. This is all nonsense, as we shall see from both the understanding of how venous incompetence really develops, and also from the multiple prevalence studies which show no link with varicose veins and weight.

However, at this point it would be worth mentioning a couple of very interesting points.

Firstly, a pelvic tumour is very frequently quoted as a cause of varicose veins. Indeed it is taught so frequently that many medical students are still told that a failure to perform a pelvic and rectal examination in the face of varicose veins is negligent. Apart from the lack of understanding of how varicose veins and venous incompetence actually develops, which will be described later in this chapter, this also shows a lack of understanding of the effect of an external pressure on the vein.

If there were a tumour growing in the pelvis and pressing on the vein from the outside, it would push the vein wall inwards from the side, restricting the flow of blood. However despite the increased resistance to blood flow that would eventually happen from this narrowing, it would not be increasing the pressure on the column of blood within the vein. As we have already seen in Chapter 2 and **figure 8**, and will be discussing again later, the pressure in a column of blood is related to the height of the column of blood alone.

Therefore, rather than causing varicose veins, a pelvic tumour would actually cause outflow obstruction and/or deep venous thrombosis. This is explained by the fact that narrowing the passage of flow in the vein increases the resistance to blood flowing out of the leg. This increased resistance means that the leg would be more full of blood (venous 'congestion') which would not result in simple varicose veins but in a general swelling of the leg. When severe enough to cause a lot of swelling, there could well be an associated increase in temperature and discolouration.

As the level of outflow resistance increases, the blood flow through the narrowed vein changes and it becomes more and more likely that the blood will clot in the vein (a deep vein thrombosis). This risk is enhanced by the fact that the presence of a malignancy also increases the risk of a thrombosis.

Therefore if a pelvic tumour is present at a size sufficient to have an effect on the iliac veins, it is much more likely to cause a deep vein thrombosis of the iliac veins and outflow obstruction of the leg (swelling, heat and discolouration), than be picked up by seeing some varicose veins in the leg.

**Pregnancy and the development of varicose veins:**

Even today, pregnancy is often quoted as a cause of varicose veins. This is usually explained as the presence of the fetus and uterus 'pushing' on the pelvic veins and 'causing pressure' on the leg veins. The observations that women going through pregnancy develop varicose veins that they did not have before pregnancy, and the fact that the pregnant uterus is such a large mass in the pelvis, has resulted in the assertion appearing logical and therefore has been unchallenged for years. However, as with the other 'obvious' causes of varicose veins due to an 'increase in pressure', this has also turned out to be wrong.

As with the story of pelvic tumours, if the problem was really due to the large mass of fetus and uterus pushing on the pelvic veins and obstructing them the result would not be the development of varicose veins; it would be the development of very swollen legs (over and above the normal swelling that we will describe next) due to the obstruction of the outflow of venous blood from the legs and the possible development of deep vein thrombosis in the iliac veins.

The actual cause of the apparent appearance of varicose veins in pregnancy is due to the fact that during the third trimester of pregnancy, between weeks 24 and 36, a normal female will increase her blood volume by 40%. It is interesting to note that she will also increase the number of red blood cells by 18% meaning that the blood is more diluted. This was often misdiagnosed as 'anaemia of pregnancy' and, before this was understood, resulted in many pregnant women having unnecessary blood transfusions. In fact, rather than being anaemic, pregnant women actually have more red blood cells. It is just that they are diluted by an even bigger increase in the surrounding fluid in the blood and so the concentration of the blood is lower.

The evolutionary reason to retain all of this extra fluid is quite sound. In normal childbirth considerable blood can be lost, particularly before modern medical techniques were around to make childbirth safer. Therefore the storage of extra blood within the mother's circulation increases the chances of survival of the mother during childbirth. The big question for us as vein specialists is where this extra blood is stored in the body.

Clearly it cannot be stored inside the head as this is already full of brain and cerebrospinal fluid. As the skull is rigid and cannot expand there is no opportunity for further storage here. Similarly it cannot be stored in the chest as the lungs have to be completely dry to enable proper gas transfer. Any increase in fluid in the chest cavity could be disastrous, causing breathing problems. The abdomen and pelvis are able to store extra blood in the visceral veins (the veins to the gut and organs of digestion such as liver) but the presence of the enlarging uterus, and the amount that the abdominal wall can extend, puts a limit on to how much can be stored.

Therefore there is a considerable amount of extra blood stored in the legs and arms. As the legs are much larger than the arms in terms of volume, they have a proportionally higher amount of extra blood. As we understand from the capillary circulation that we looked at in Chapter 2, an increase in the amount of blood will also increase the amount of tissue fluid, hence the swelling of the limbs in pregnancy.

If the body suddenly retained 40% more blood volume this would have dramatic physiological effects, and the veins in particular would not be able to simply dilate passively. However the beauty of biology is the coordination of many different processes to reach an endpoint. For a baby to be born through the rigid bony ring of the female pelvis, it is necessary for the pelvis to be allowed to expand a little so that the head does not get stuck. Therefore the ligaments holding the three bones of the pelvis together (the pubic symphysis and the two sacroiliac joints) are all softened by the pregnancy hormones (oestrogen and progesterone), allowing the joints to widen and the rigid bony ring to open a little, assisting childbirth.

This is one of the major reasons for low backache in later pregnancy. The large triangular bone at the base of the spine, the sacrum, bears the weight of the body and head via the vertebral (spinal) column. The sacrum is usually supported by the two iliac bones which transfer the weight around the pelvic bones and into the legs. However during pregnancy the softening of the ligaments of the sacroiliac joints by pregnancy hormones relax this support, meaning that the ligaments get stretched, the joint opens up and the bones move, stretching the joint and causing pain.

The high levels of oestrogen and progesterone also affect the venous walls, dilating all of the veins and allowing the extra 40% of blood to be accommodated without any major increase in pressure within the vein lumen.

Therefore any woman who already has underlying varicose veins which are too small to be seen before pregnancy, will appear to develop varicose veins during pregnancy as these veins enlarge under the influence of the oestrogen and progesterone and are filled by the extra blood.

Although this appears counter intuitive, research has confirmed that this is the case. In the 1990s a study was performed in Chester, in the UK, that examined the leg veins in women having IVF (In-

Vitro Fertilization) using a duplex ultrasound. These women were then followed through their pregnancy.

All of the women who appeared to develop varicose veins during pregnancy were found to have already had problems with their valves, often unbeknown to them but picked up on the duplex ultrasound at the initial scan.

More evidence is found in the prevalence studies which have been performed into who gets varicose veins. Although varicose veins were thought to be predominantly a female problem in the past, this erroneous conclusion has been shown to be due to poor study design. In these studies, doctors would merely report who came to them complaining of varicose veins - and we know that men rarely go to doctors unless they have to. When studies were performed correcting this selection bias and the researchers went out into the community, such as was done in Edinburgh in the late 1990s, and people were stopped and examined with Duplex Ultrasound at random, it was found that the prevalence of varicose veins is almost identical between the sexes.

If pregnancy really did cause varicose veins, we would see a much bigger prevalence in women than men, and this prevalence would be found in the childbearing years, mainly between 20 to 40. However, this is not found.

Therefore pregnancy does not cause varicose veins or venous reflux. However, if the valves have already stopped working and venous reflux is already present, pregnancy will make varicose veins more apparent and will worsen the clinical appearance.

One final note of caution: the whole argument about pregnancy outlined above is correct in women with varicose veins and venous reflux restricted to their legs only. There is a very interesting condition called 'pelvic congestion syndrome', which is the result of varicose veins inside the pelvis. Research at The Whiteley Clinic has shown that if a woman with pelvic congestion syndrome undergoes vaginal birth (normal delivery), there is a high incidence of the pelvic varicose veins communicating through the wall of the vagina and vulva and into the veins of the legs.

In this situation, the development of varicose veins in the vulva and vagina, often leading into varicose veins in the upper thigh and into the leg, is totally caused by the birth of the child. Although this is a special case there are some parallels with the story of leg varicose veins in pregnancy above. For this pelvic reflux into vaginal, vulval and leg varicose veins to occur, there is a requirement for pelvic varicose veins to be present as pelvic congestion syndrome pre-delivery. Thus even in this special case, pregnancy and delivery isn't the only cause of these veins. We will return to this later in this book, but will discuss the condition at length in a subsequent clinical book in this series.

## Ascending reflux

One of the most formative moments of my life in understanding venous reflux disease and varicose vein formation was when I learnt how to perform duplex ultrasound examinations myself. I started performing duplex ultrasound as a part of my research as a junior doctor in 1992. Although much of my early experience was looking at arteries, I became very interested in the function of veins, particularly as this new technique could show the blood flow in real-time.

By 1994, I had realised that many of the people who presented with varicose veins due to Great Saphenous Vein incompetence actually had competent valves that were still working at the saphenofemoral junction, and also quite often in the upper few centimetres of the Great Saphenous Vein. This is of course impossible if the domino theory of venous reflux was correct.

I tried to discuss this on several occasions with vascular surgeons. Unfortunately vascular surgeons tend to be more interested in arterial disease, which occupies the majority of their time. As duplex ultrasound was also a new investigation, I was usually told that the observation was just wrong and the duplex scan must have been performed badly. In 1998 when I set up my own vascular unit I discussed this with Judy Holdstock, a highly skilled vascular technologist, who has been involved in virtually all of my research. She was most unimpressed at the observation, saying that she had seen it frequently herself. When I asked her why she didn't report this, her response was that she had learned that surgeons knew what they wanted to find, and would not like reports that pointed out a complexity if it did not change the treatment they were going to perform. Therefore she had learned to report what the surgeons understood rather than the precise and accurate observations that she was making.

Between 1998 and 2000, we performed over 1000 duplex ultrasound scans on people with varicose veins and, as part of our routine examination, we performed a duplex ultrasound scan on the veins of the other leg, i.e. the leg without clinical varicose veins (**Figure 22**).

Unsurprisingly, in a large number of these 'normal' legs, early varicose veins or venous reflux could be found. This gave us insight into the sequence of events that occurs in the development of varicose veins. In all of the legs that were developing varicose veins from the Great Saphenous Vein, in which some of the valves were

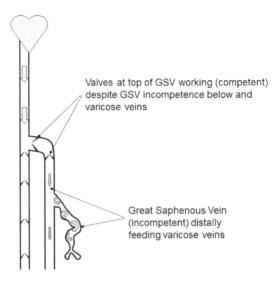

Valves at top of GSV working (competent) despite GSV incompetence below and varicose veins

Great Saphenous Vein (incompetent) distally feeding varicose veins

Figure 22: 'Ascending progression of reflux' - GSV reflux and varicose veins beneath competent valves – shown to be the cause of reflux by us in 2001

still competent and working normally in the GSV, we found a similar pattern: the scan showed competent valves still working at the saphenofemoral junction, and many of them also had competent and working valves just below this point. We did not find a single case where a GSV with an incompetent valve at the saphenofemoral junction, allowing blood to reflux, which then had normal competent valves below this point.

This clearly showed that the domino theory was completely wrong and the actual sequence of valve failure in venous incompetence is exactly the reverse of what it suggested. We found that early incompetence of the valves seems to start lower in the leg, sometimes around the knee or often just below or above that level, and usually in a tributary of the Great Saphenous Vein. This then seems to corrupt the Great Saphenous Vein at that level, causing reflux in the trunk of the Great Saphenous Vein at just that point. With further time, the next valve up becomes incompetent and fails, followed by the one above that, resulting in an ascending pattern of reflux. We published our findings in 2001.

Although we now know the pattern of how this venous reflux develops and the order in which the valves become incompetent, we have not been able to identify the cause as to exactly why it occurs this way. It may be that as the veins dilate, the valves above the area of dilatation do not meet any more and fail, allowing more blood to reflux and thus for the dilatation to increase further. However, as there is clearly a genetic preponderance to the development of varicose veins, the reason may be more complex and might not just be structural. This will be an area for further research in the future.

### Phase 1 and Phase 2 reflux of Great Saphenous Vein varicose veins (passive reflux)

This observation suggesting ascending reflux, where the top valve of the Great Saphenous Vein is still competent at the saphenofemoral junction whilst having incompetent valves lower in the Great Saphenous Vein (causing reflux and in some cases varicose veins), gave rise to several problems for traditional surgeons.

Firstly, when we initially presented this work, we were not believed. One of the objections raised was that if the top valve was working, blood would not be able to get into the Great Saphenous Vein to be able to reflux down it. We were told it was like putting 'a top on the bottle' preventing any blood from refluxing into the vein from above, and therefore stopping any blood from refluxing down the vein.

However when careful duplex ultrasonography is performed, the source of the blood that is seen to reflux down the Great Saphenous Veins, despite a competent valve at the saphenofemoral junction, becomes obvious (**Figure 23**).

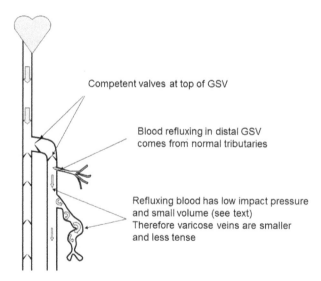

Figure 23: 'Phase 1 passive reflux' – varicose veins present despite competent valves at top of GSV

In the normal venous system of the leg, venules feed into tributaries, which then feed into the truncal veins and from there into the deep veins (see Chapter 1). In the situation that we are discussing, where the valve at the top of the Great Saphenous Vein at the saphenofemoral junction is competent but the valves below this point are incompetent, normal tributaries are still feeding blood into the Great Saphenous Vein. However instead of this blood being pumped upwards into the deep vein, the valve failure in the lower Great Saphenous Vein allows passive reflux of this blood downwards due to gravity.

Secondly, if this reflux occurs in the Great Saphenous Vein despite a competent valve at the saphenofemoral junction, it instantly shows that the Trendelenburg operation (a simple ligation of the Great Saphenous Vein at this junction) will have no effect at all on this reflux.

However, we do know that in some cases where all of the valves have become incompetent in the Great Saphenous Vein (including the top valve at the saphenofemoral junction), the Trendelenburg operation does improve matters (although subsequent research has shown that it is an incomplete operation with high recurrence rates for reasons that will be discussed in a later book in this series) (**Figure 24**).

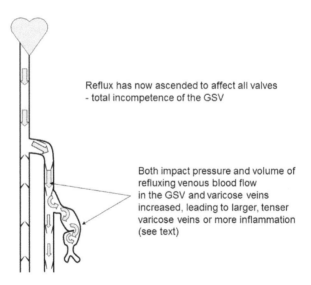

Reflux has now ascended to affect all valves - total incompetence of the GSV

Both impact pressure and volume of refluxing venous blood flow in the GSV and varicose veins increased, leading to larger, tenser varicose veins or more inflammation (see text)

Figure 24: 'Phase 2 passive reflux' – all of the valves in the truncal GSV become incompetent

It therefore became clear that firstly, varicose veins and venous reflux develops in an ascending pattern. Secondly, the passive reflux of the Great Saphenous Vein can be divided into two distinct phases:

Phase 1 reflux - when there is a competent valve at the saphenofemoral junction but there is reflux in the Great Saphenous Vein (**Figure 23**).

Phase 2 reflux - when the reflux has ascended and affected the topmost valve of the Great Saphenous Vein, allowing reflux of the whole column of blood from the inferior vena cava and iliac veins to flow into the incompetent Great Saphenous Vein (**Figure 24**).

In Phase 1 reflux, the volume of blood refluxing at any time down the Great Saphenous Vein is low as it is only the blood fed into the vein from the tributaries. As it is falling from a level below the groin, the 'impact pressure' of that blood hitting the vein wall at the ankle or sidewalls in varicose veins is low. Therefore, varicose veins seen in Phase 1 reflux tend to be small and often not very tense or tender.

In Phase 2 reflux, there is a much larger volume of blood refluxing at any time down the Great Saphenous Vein. As all of the valves in the Great Saphenous Vein have gone, the blood available to reflux into this vein is the whole volume of blood in the inferior vena cava and iliac veins as well as any blood supplied by the tributaries.

In addition, as the blood is now falling from a height of the right atrium of the heart, the 'impact pressure' of the blood hitting the vein walls at the ankle or in the varicose veins is higher. As a result of both of these factors, varicose veins associated with Phase 2 reflux tend to be bigger, more tense and more often tender. In addition there is likely to be more inflammation in the tissue around the ankle, which will be discussed later.

You will note that I have been very careful explaining the difference between phase 1 and phase 2 reflux as passive reflux of the Great Saphenous Vein. The reason for this will become apparent later in the book when we discuss active reflux, which is completely different in nature and does not have different phases.

Similarly, I have also been very careful in explaining the differences between phase 1 and phase 2 in terms of volume flow of refluxing venous blood and 'impact pressure'. This is not just pressure of blood or 'venous hypertension' that has commonly been talked and written about in the past. 'Venous hypertension' is a misnomer, which encourages an erroneous view of venous reflux disease and chronic venous insufficiency. As such, the term 'venous hypertension' should never be used in venous reflux disease.

# CHAPTER 6

# 'Venous hypertension' - an obsolete term

'Venous hypertension' literally means abnormally high blood pressure (hypertension) in the veins.

I suppose it would appear logical to assume that all of the problems due to venous reflux disease (chronic venous insufficiency) are due to high pressure of the venous blood at the ankles. After all, when someone has varicose veins and they are standing up, the veins appear to be tense. Moreover, if the varicose veins bleed when the person is standing up, the venous blood comes out of the ruptured varicosities at a very high pressure.

Unfortunately, like so many things in the understanding of venous problems, this is not the case. Rather shockingly for most doctors and nurses, 'venous hypertension' does not exist in venous leg disease - certainly not in the way it is thought of traditionally.

To understand this fully, we will go through a series of scenarios which when put together, will lead us to understand why 'venous hypertension' should be an obsolete term when discussing varicose veins, venous eczema or leg ulcers due to venous reflux disease of venous incompetence.

## Who has the worst 'venous hypertension'?

This is a question I have asked every delegate on all of my courses over the last decade and at the time of writing this book, not a single person has got it correct.

Look at **Figure 25**.

These matchstick men represent three people, all of whom are the same height and all are standing stationary.

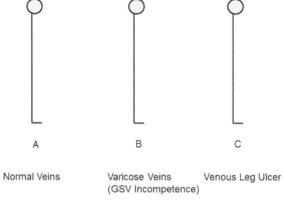

A         B         C

Normal Veins     Varicose Veins     Venous Leg Ulcer
(GSV Incompetence)

Three subjects, all of the same height, A = normal veins, B = varicose veins, C = venous leg ulcer.
Which has the highest venous pressure at the ankle (ie: venous hypertension)?

Figure 25: Who has 'venous hypertension'?

- The first person, A, has normal veins both clinically and also when tested using duplex ultrasound.

- The second person, B, has total incompetence of their Great Saphenous Vein and varicose veins arising from the Great Saphenous Vein.

- The third person, C, has a venous leg ulcer.

If we were to measure the pressure (i.e. measure the 'venous hypertension') in a vein on the foot, or at the ankle, who would have the highest pressure? And who would have the lowest?

Surprising though it may seem at first instance, they all have the same venous pressure at the ankle. The pressure in the veins at the foot or ankle (or the measured 'venous hypertension') bears no relationship at all with the clinical appearance or even the presence of venous reflux disease (chronic venous incompetence).

This is usually a revelation for people, even if they are dealing with patients with thread veins, varicose veins, venous eczema or leg ulcers on a daily basis. The explanation, however, is obvious when we ignore the medical mumbo-jumbo we have been brought up with and we think clearly from first principles.

## Measuring pressure in a column of fluid

To understand this fully, we need to return to some very basic physics that we were probably taught at school and we talked about in Chapter 2.

We have to consider a column of fluid standing on a surface (**Figure 26**). The column of fluid exerts pressure on the surface on which it is standing.

The downward force exerted by the column of fluid is its weight. The weight of the column of fluid is due to 3 factors:

• How much of the fluid is present

• The density of the fluid (i.e. the mass of a set volume - usually one ml for fluids)

• The force of gravity (or more precisely the acceleration due to gravity)

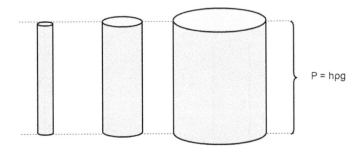

$P = h\rho g$

As the density of the fluid and gravity are the same for all 3 columns, the TOTAL WEIGHT of the columns increases with size (as there is more fluid)
**BUT**
The PRESSURE (which is the force (or weight) per unit area) is the same for each column. The wider the column, the larger the area.

Figure 26: Measuring the pressure of a column of fluid and comparison with the weight of the column

In physics we have to identify these points to make sure we are calculating things precisely, allowing us to predict changes if any of the individual things that we are studying change; for example, if we wanted to think about the pressure on the moon where gravity is less than that on Earth, or if we wanted to think about the effects of a less dense or more dense fluid.

However as we are thinking about humans on earth, the calculations are much easier for us. As blood has almost identical density in different humans and gravity on earth is virtually the same wherever we are, these two parts of the three part calculation are constant and do not change.

If we were considering the weight of the column of fluid we would need to know the dimensions of the column, i.e. the height and the circumference, so that we would know the total volume of fluid and therefore would know the total weight. However, we are not calculating weight but rather pressure.

Pressure (P) is the downward force (F) divided by the area (A) it is spread over:

$$P = F/A$$

This means that in a column of blood where we know that the density and gravity is going to be constant, and we are only thinking about how much force is being placed on one set area of the surface that it is standing on, the only thing that can be varied is the height of the column (**Figure 27**).

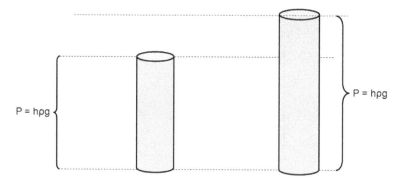

In columns of different heights, the taller the column, the more mass ("weight") there is above each part of the area it is standing on.
Therefore the pressure (force per unit area) INCREASES the HIGHER the column is and so both weight of column and pressure increase

Figure 27: Measuring the pressure of a column of fluid and seeing how it changes with height of the column

Therefore, regardless of circumference of the column of fluid, the higher the column of blood, the more pressure it produces at its base and the lower the column, the lower the pressure.

### Relating pressure in a column of fluid to 'venous hypertension'

If we now return back to our original three subjects, we can see that they are all the same height (**Figure 28**).

As alluded to earlier in this book, it does not matter how tall the subjects are when we measure pressure, but how high their hearts are above their feet. This is because all haemodynamic pressures measured within the body are measured with reference to the heart. If subjects A, B and C are all the same height, then it is likely that

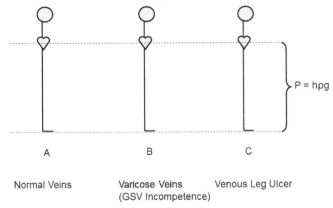

$$P = h\rho g$$

A        B        C

Normal Veins     Varicose Veins     Venous Leg Ulcer
(GSV Incompetence)

All three subjects have THE SAME pressure at the ankle which is calculated by pressure = hρg where h = height of column of blood from heart to ankle, ρ = density of blood and g = gravity

Figure 28: Explanation why 'venous hypertension' doesn't actually exist

their hearts would also all be the same height above their feet.

When they are at rest and standing blood has been pumped by the heart through the arteries, through the capillaries and has gone into the venous circulation, causing a column of blood to extend from foot to heart. Provided they do not move and activate any part of the venous pump, these three columns of blood will all be exactly the same height, and will therefore cause exactly the same pressure to be present in the ankle and at the foot.

This theory has been observed and measured by many studies and has been shown to be correct.

However, when we put this together with our clinical knowledge of patients with valves that are incompetent, who also have varicose veins and leg ulcers, it seems to be nonsense. Therefore we need to explore what is the actual cause of the damage to the veins that gives rise to varicose veins, venous ulcers and all the other sequaelae of venous reflux disease (chronic venous incompetence), and this is what we will start doing in the next chapter.

## Summary of 'venous hypertension'

'Venous hypertension' as commonly understood when discussing varicose veins, venous reflux disease or venous incompetence does not exist. All patients at the same height have the same venous pressure at the ankles when standing stationary, regardless of the presence and severity of any venous abnormalities such as varicose veins or leg ulcers.

The cause of the damage that goes on to cause these problems is therefore not venous hypertension.

# CHAPTER 7

# Inflammation in venous reflux disease (chronic venous insufficiency)

So, now that we have discovered that 'venous hypertension' doesn't exist, it leaves us in an uneasy position.

We can clearly see people who have varicose veins, problems with the skin around their ankles that are leading towards venous leg ulcers (such as venous eczema, lipodermatosclerosis - LDS, haemosiderin), or even with a venous leg ulcer; and when we perform any specialist tests (such as a venous duplex ultrasound) we can see that the valves in some of their leg veins have lost their function and are incompetent, allowing blood to reflux down the veins.

Furthermore, if we treat the venous reflux disease properly in these patients, making sure that the blood does not reflux down these veins any more, the clinical conditions get better; venous leg ulcers heal, skin changes reverse and improve, varicose veins are cured.

Thus, there is a very clear link between valve failure causing venous reflux disease, and this reflux of venous blood giving rise to the clinical damage that we can see lower in the leg.

For over one hundred years, doctors and nurses have been able to safely hide behind the explanation that these clinical changes are due to high venous pressure caused by the varicose veins or venous reflux i.e. venous hypertension. Having taken away this explanation, we now have to identify the actual cause.

### Back to our model subjects - A, B and C

To understand the damage that causes the majority of the problems in venous reflux disease, we are going to our three model subjects that we used previously - subject A, B and C (**Figure 29**).

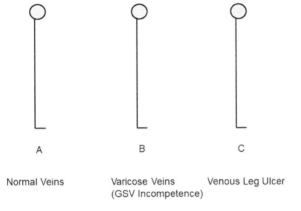

A                    B                    C

Normal Veins         Varicose Veins       Venous Leg Ulcer
                     (GSV Incompetence)

Figure 29: Measuring inflammation in venous reflux disease – the subjects

As before:

- Subject A has normal veins both clinically and also when tested with duplex ultrasound

- Subject B has total incompetence of the Great Saphenous Vein allowing phase 2 passive reflux from heart to ankle

- Subject C has a venous leg ulcer (which is due to truncal incompetence of the Great Saphenous Vein)

As we know from the previous chapter, all of these subjects are the same height and therefore have the same venous pressure at the foot and ankle.

## Observing inflammation from Phase 2 passive reflux

Although we have already shown that all 3 subjects have the same venous pressure at the foot and ankle, this is only the case after the subjects have been standing stationary for some time.

In the 1980s, a research group in the Middlesex Hospital in London, UK, developed a very simple stimulatory test to observe what happens with passive reflux due to gravity (**Figure 30**).

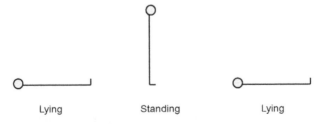

Lying              Standing              Lying

Figure 30: Test to observe effects of phase 2 passive reflux in the veins

In this test subjects were asked to lie down for a while to stabilise the pressure changes within the circulatory system, and also to clear out any inflammatory processes that might have occurred previous to lying down. They would then be asked to stand up quickly and to stay standing very still, so as not to activate the vein pump, until the pressures in the circulatory system had once again stabilised.

Once stabile, the subject was then asked to lie down again and was once again allowed to stabilise for up to 10 minutes to allow any circulatory pressure changes or inflammatory changes to settle.

Using this simple system the research team then performed several tests including taking blood samples from the foot and ankle when lying, when standing, and then when lying again. It is the results of these tests that indicate how damage is caused in the lower leg by valvular incompetence in the veins and venous reflux disease.

### Subject A - inflammatory changes at the ankle with no passive venous reflux

We will look first at the results from subject A, the person with normal legs and normally functioning valves in all of the leg veins (**Figure 31**).

When blood tests looking for markers of inflammation are taken from the blood at the ankle or foot, the levels of most of the inflammatory markers (such as activated white blood cells) are normal when lying, remain normal when standing, and then stay normal when lying back down again.

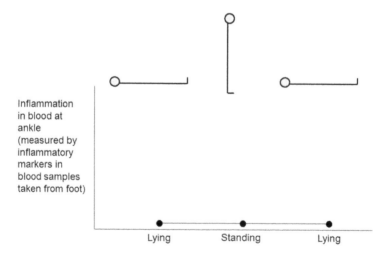

Figure 31: Results of inflammatory markers on standing in subject A - normal veins in legs with functioning valves

This shows that in a person with no venous reflux, and with valves that are working within the venous system of the legs, that standing does not cause any inflammatory reaction in the veins at the foot or ankle; i.e. the normal venous pressure at the ankle on standing does not cause inflammation.

## Subject C - inflammatory changes at the ankle with the venous leg ulcer due to phase 2 passive reflux in the GSV

The results from subject A can be readily contrasted with the results from subject C who has venous leg ulcers and phase 2 passive venous reflux due to incompetent valves in the whole of the GSV (**Figure 32**).

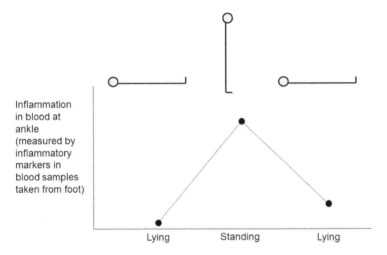

Figure 32: Results of inflammatory markers on standing in subject C - venous leg ulcer in legs with non-functioning valves

In this subject, the inflammatory markers are normal after the patient has been lying for some considerable time, but on standing there is a marked rise in all of the acute inflammatory markers. On lying down again the inflammatory markers start returning to normal and, depending how long the patient has been lying, the more normal they become.

What we can conclude from comparing these two subjects is that something is happening in subject C on standing, which is causing the venous leg ulcer. It cannot be 'venous hypertension', as we have already seen in Chapter 6 that both of these subjects have got the same pressure at the ankle on standing. However, it is clear that having phase 2 passive reflux in a Great Saphenous Vein is causing an increased acute inflammation on standing in the lower leg.

It is well known within medicine, and also in normal life, that chronically repeated acute inflammation in one area ends up causing tissue damage if allowed to continue. For instance, repeated acute inflammation in arthritis results in the eventual destruction of joint cartilage and eventually the joint itself. Similarly, anyone who has an item of clothing or equipment they have to wear or use regularly, that rubs in one particular place, will end up with a sore area that is red in the first instance (acute inflammation); and if the same conditions continue, the rubbing will eventually cause either thickening of the skin, or discolouration and scarring of the skin that may even cause a dimple (chronically repeated acute inflammation).

Therefore, from the findings outlined above and in **Figure 32**, we can see there is an association between acute inflammation at the ankle and standing in a person with phase 2 passive reflux. We can also see that there is an association between chronically repeated acute inflammation and skin damage – that we know clinically precedes venous leg ulceration.

We should now look at the middle subject to glean further information from these observations.

### Subject B – inflammatory changes at the ankle with varicose veins phase 2 passive reflux in the GSV

The results from subject B are outlined in **Figure 33**. In this

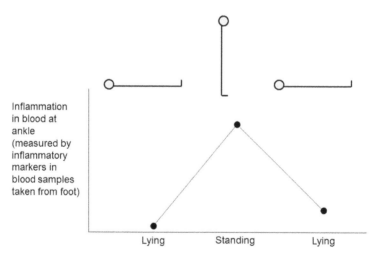

Figure 33: Results of inflammatory markers on standing in subject B - varicose veins in legs with incompetent valves in GSV (phase 2 passive reflux) (NB: This is only true when the 'silencing' effect of the varicose veins has been overcome – see later in text)

subject, as with the others when they are lying down, there is no acute inflammation at the ankle. However on standing it is rather surprising to note that the acute inflammation is usually more or less at the same level as subject C who has the venous leg ulcer, despite the subject being better clinically as they only have varicose veins and not a leg ulcer. On lying down subsequently, the acute inflammation settles with time as in subject C.

Although one might expect that a subject with varicose veins - due to phase 2 passive reflux in the Great Saphenous Vein - would have a less severe level of acute inflammation on standing than the subject with the venous leg ulcer, this is purely because we are assessing the clinical appearance and not the underlying cause. The clinical appearance of varicose veins (subject B) and venous leg ulcer (subject C) would appear to an uninformed observer to be completely different. This is why doctors and nurses in the past - and unfortunately many even today - state that varicose veins are 'cosmetic only' and do not need treatment, whereas they would never make the same statement about the venous leg ulcer.

As we can see, the underlying problem for both subject B and subject C are identical: both have phase 2 passive reflux in a Great Saphenous Vein. Furthermore, both have virtually identical inflammation profiles on measurement of acute inflammation at the ankle. So why haven't subject B and subject C got the same clinical outcome for the same underlying cause?

There are two reasons for this, the first will be explained in the forthcoming chapter and the second is that they are almost identical conditions, separated by the length of time they have had the condition – i.e. the amount of time the recurrent acute inflammation has been allowed to affect the skin.

If subject B is not treated, and the acute inflammation is allowed to continue, the damage of the recurrent bouts of acute inflammation will build eventually causing the inflammatory changes at the ankle that we see clinically: initially swelling and tenderness, and then as the condition worsens venous eczema, lipodermatosclerosis, brown staining (haemosiderin deposition) and eventually breaking down of the skin - venous ulceration.

## Summary

At this stage in our investigation as to what is going on in the leg with passive venous reflux due to valve failure, we have answered many questions but these in turn have posed further questions.

We now know that passive venous reflux in the Great Saphenous Vein starts lower in the leg as phase 1 reflux, and then proceeds to phase 2 reflux as the valvular incompetence ascends to the saphenofemoral junction.

Once this phase 2 reflux has become established it results in inflammation at the ankle on standing. However, we don't have a clear understanding at this point as to why this same inflammation isn't found in the subject who has normal valves (subject A). This subject has the same pressure at the ankle on standing as subject B & C, as they are all the same height. We have also partially understood why two people, subject B and subject C, can have the same phase 2 passive reflux in a Great Saphenous Vein but have completely different clinical presentations. The reason we have identified is the time factor of how long the recurrent bouts of acute inflammation have been present at the ankle. A second reason will be explored in the next chapter.

In Chapter 8 we will clear up most of these points and will pull together most of the concepts we have discussed so far.

# Pressure, flow and inflammation - how they relate in venous reflux disease

In this chapter we are going to bring together all of the concepts and information that we have examined up to this point. Hopefully, by the end of it, you will have an 'Eureka' moment and will be well on your way to understanding the basics of venous reflux disease. Without doubt, if you do reach that point, it will have a significant impact on your attitude to what you thought you knew before. If you are treating veins in any capacity, it will make you completely re-assess your current practice. And if you don't treat veins but have an interest in the subject, you will start to understand why so many people get sub-standard investigations and treatment for venous conditions.

However, once you do reach that point, there is still an awful lot to learn about veins and venous treatments - but it will be from the basis of a firm foundation of knowledge.

So let us get on with pulling together how venous pressure, blood flow and inflammation interact to cause venous reflux disease. For this discussion we are going to change to 2 new subjects, subject X and subject Y (**Figure 34**).

As with our previous subjects, X and Y are both the same height, meaning that both of them have the same pressure on passive standing at the ankles. Subject X has normal veins with all of the venous valves working normally. Subject Y has phase 2 passive reflux in the Great Saphenous Vein; i.e. total incompetence of the Great Saphenous Vein meaning that venous blood can reflux from the right atrium to the ankle on standing.

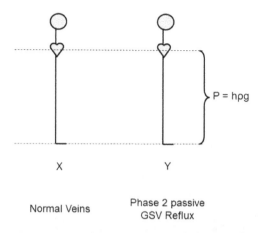

X                          Y

Normal Veins          Phase 2 passive
                      GSV Reflux

Figure 34: The pressure at the ankle in two people of the same height standing still – one with normal veins, one with phase 2 passive reflux down the incompetent GSV

### Pressure changes on moving from lying to standing between a subject with normal veins and one with phase 2 passive reflux

To see the changes that we need to, we must stimulate the venous system in a way that emulates the natural processes of everyday life. Therefore we are going to use the first half of the test used in the last chapter: simply lying the patient down and letting them settle for approximately 10 min, and then standing up stationary so that no vein pump in the leg is activated (**Figure 35**).

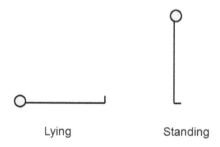

Lying                    Standing

Figure 35: Model of how to measure rate of change of pressure in the veins at the ankle

### Subject X – normal veins and valves

The first of our subjects, subject X, is represented in **Figure 36**. During this process the pressure is being monitored in a vein on the foot. When lying down, the pressure is approximately 15 mmHg.

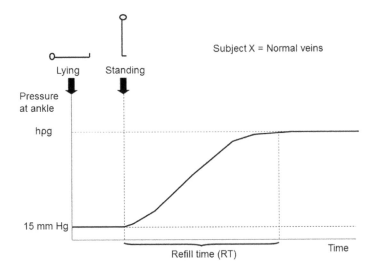

Figure 36: Graph showing the rate of change of pressure in the veins at the ankle

This is the pressure of the blood exiting the post-capillary venules as we discussed in Chapter 2. This represents the potential energy to drive the blood back through the venous system to the heart, which has a venous pressure in the right atrium of approximately 0 mmHg.

On standing, the pressure at the foot rises to the maximum, which depends on the subject's height as discussed in Chapter 6. This can be measured as hpg.

The important thing to notice from this graph is that there is a gentle increase in pressure from the baseline 15 mmHg to the standing pressure of hpg. We measure the time taken to reach this pressure from standing and call it the refill time (RT).

The refill time in normal people is usually somewhere between 25 and 40 seconds. The reason it takes this length of time can be understood if one considers the circulation.

On standing, the column of blood in the inferior vena cava and pelvic veins starts to reflux due to gravity, which closes the valves at the top of the deep vein (femoral vein) and the top of the Great Saphenous Vein (at the saphenofemoral junction). Therefore, for the complete column of blood from ankle to heart to reach the pressure of hpg, blood has to fill the whole of the venous system below these valves to make one continuous column of venous blood. The veins below the top valves in the leg fill through the normal circulation. The heart pumps blood through the arteries, into the capillaries where it travels into venules and then into these leg veins.

As the blood slowly fills the veins through this normal route, the column of blood accumulates slowly resulting in the slow rise of pressure and the refill time being over 25 seconds.

Now if you have understood all of the arguments in this book so far, you will note a problem here. In Chapter 4 it was pointed out that the venous pressure at the foot and ankle was only 15 mmHg, which would only get the venous blood up the veins to the lower leg below the calf – nowhere near enough to create the column that we need from the valves at the top of the leg in both saphenofemoral junction and the vein to ankle.

However if you have also understood the venous circulation fully, you will realise that all of the tissue of the lower leg, upper leg and in fact the whole body needs nutrients and oxygen and therefore has capillary networks feeding them. Therefore venous blood is emerging from capillaries into venules at 15 mmHg in the calf, around the knee, in the thigh, etc and all of this blood is flowing into the deep veins and the venous trunks of the leg, including the Great Saphenous Vein. Therefore, although it suited us from a pressure point of view to only consider blood at the ankle before, when it comes to this filling of the column of blood when the subject is standing still, we have to think about how the system is working in reality.

This sudden exposure to reality does not change any of the arguments in this book up to this point. Although venous tributaries are feeding these veins in the legs all at 15 mmHg, this does not represent any driving pressure that would return venous blood to the heart when the leg is stationary. Indeed, the blood flowing from the tributaries into the veins would simply reflux down to the next valve, where it would stay until the venous pumps ejects it under pressure towards the heart.

### Subject Y – Phase 2 passive reflux in the GSV

The pressure trace for subject Y measured in a vein at the foot is demonstrated in the same format as the previous subject in **Figure 37**.

Subject Y has phase 2 passive reflux in the Great Saphenous Vein, i.e. none of the valves in the Great Saphenous Vein are working allowing reflux from right atrium of the heart to the ankle.

When lying down, just as with subject X, the pressure of the blood exiting the capillaries into the venules is 15mmHg. On standing, the pressure rises briskly to the same pressure as in subject X because both subjects are the same height.

However in subject Y the refill time is dramatically less, typically between two and ten seconds. This difference is due to the venous blood in the inferior vena cava and pelvic veins refluxing, by gravity

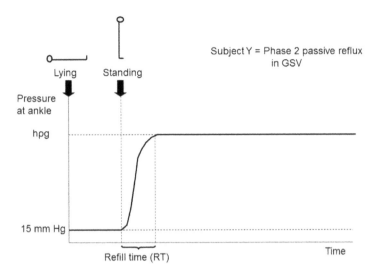

Figure 37: Graph showing the rate of change of pressure in the veins at the ankle in patient with phase 2 passive reflux in the GSV

on standing, down the leg towards the foot and ankle. In this case the valves in the Great Saphenous Vein are all incompetent and so the blood continues refluxing down the Great Saphenous Vein to the ankle, with nothing to prevent this backward flow of blood on the way.

### Rate of change of venous pressure at ankle and reverse venous flow (venous reflux)

When these two pressure profiles are compared directly, the difference becomes obvious (**Figure 38**).

The pressure in the venules (or in the veins at the foot and ankles) when lying down is the same in both of the subjects, as is the pressure when both are standing. The difference between the two subjects is how quickly the pressure accumulates i.e. the rate of change of pressure.

This is the key to venous reflux disease and almost all of the problems that were previously attributed to 'venous hypertension'. It once again confirms our conclusion in Chapter 6, that there is no such thing as 'venous hypertension'. What it shows us is that with phase 2 passive reflux in the Great Saphenous Vein (which in Chapter 7 we concluded was associated with varicose veins and inflammation in the lower leg) there is a much increased rate of change of pressure at the ankle on standing.

As we have discussed in the section above, this increased rate of

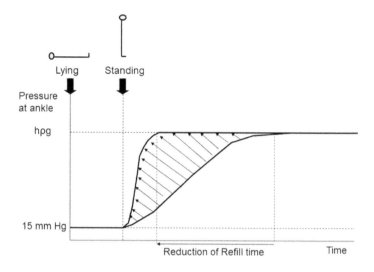

Figure 38: Graph comparing the pressure profiles between a person with normal veins and one with phase 2 passive reflux in the GSV

change of pressure at the ankle is due to an increased reverse flow of blood (or venous reflux) down the incompetent Great Saphenous Vein.

### Inflammation and rate of change of pressure at the ankle (impact pressure)

In Chapter 7, we discussed that phase 2 passive reflux down the Great Saphenous Vein was associated with acute inflammation at the ankle and foot. We found this by noting that the subject A, who did not have this reflux, did not have any acute inflammation measurable on standing.

This can be likened to many areas of medicine where a certain action performed slowly will not cause damage whereas the same action performed with the same total force but far faster will cause damage. Examples of this include birth, where a baby coming through cervix and vagina slowly rarely causes damage, but the faster the baby is born, the more likely is the ripping of cervix or vagina; or opening the chest in thoracic surgery where opening the ribs slowly allows good access to the heart and lungs without any damage whereas pulling the ribs apart too quickly results in fractures of the ribs.

Therefore, we can conclude that it is the large volume of blood refluxing down the Great Saphenous Vein to the ankle causing the sudden rise of pressure that is causing the acute inflammation.

This isn't really very surprising. If we were to drop water into a bath from the same height as our heart is to our feet, we would see considerable bubbling and splashes. This represents the transfer of the potential energy of blood at a level of the heart falling and hitting the veins at the ankle, releasing kinetic energy. It is been likened to being kicked in the lower ankle, which also causes inflammation - the only difference being that the kick is external whereas the trauma from the refluxing blood is internal.

I have always found it useful to think of this as 'impact pressure' as it is both a combination of the increased pressure but also an effect of the increased volume of blood hitting the veins at the bottom of the leg and forcing them to expand.

## Inflammation in venous disease

As to exactly how this inflammation occurs and exactly how the process of inflammation has been triggered off has been a matter for a lot of research. However, although this is of interest to researchers and there is a possibility it may spawn some interesting therapies in the future, for our understanding here we must just see the principles. We see that this blood refluxing quickly down the Great Saphenous Vein causes a sudden increase in pressure and in volume in the major veins in the ankles, acutely stretching the walls. As these major veins are fed by smaller veins, venules and capillaries, this wave of pressure and blood flowing backwards has the same effect on each of these structures, also disrupting the delicate balance of the venous end of the capillaries in the area.

This sudden change of pressure and volume causes inflammation. Inflammation is characterised by four signs:

- Redness - due to dilatation of the local capillaries

- Pain - due to the release of certain chemicals to attract white cells to help clear the inflammation and start healing

- Swelling - due to the release of fluid from the damaged capillaries and small venules

- Heat - due to dilatation of the local capillaries

It is not surprising then, that in advanced venous reflux disease, we see patients with swollen ankles, which are tender, red and warm to the touch.

### Why do we not see this early in venous reflux disease?

Once the principles above have been understood, it is important to remember that as with all things in medicine, the arrival at this stage has been a process. Initially in phase 1 reflux, there was very little impact pressure due to both low volume of blood refluxing down the vein and also the lower total pressure, as the height of the column of fluid was only to the lowest competent valves.

Even when phase 2 passive reflux starts, the main Great Saphenous Vein is relatively narrow at the beginning of phase 2 passive reflux, giving some resistance to the refluxing blood, thereby reducing the flow and impact pressure. Inflammation that does start accruing at this stage is around the capillaries and venules deep under the skin, which can be seen microscopically in samples cut out in research studies. However the inflammatory changes are not noticeable clinically to those who are looking from outside of the leg.

As phase 2 passive reflux worsens the Great Saphenous Vein starts dilating allowing larger volumes of blood to reflux, with less resistance to the reflux, therefore increasing the impact pressure at the ankle on standing. The increased inflammation once again is centered around the veins and capillaries, and so it can be many years before the damage is sufficient to have spread through the subcutaneous tissues, to the dermis and be visible externally.

It should be noted here that the same arguments apply to those few people who have total incompetence of their deep venous system. The reflux is the same, but the 'impact pressure' worsens quicker due to the larger diameter of the deep veins and the larger volume of refluxing blood.

Thus one of the most important things to learn from this book, is that the assessment of severity of venous reflux disease cannot be made by looking at the outside of the leg. It is a great pity that almost every doctor, nurse or healthcare professional still does not understand this and patients are regularly told that their varicose veins, or other leg conditions due to venous reflux, are 'not serious' and do not warrant referral to a venous expert for investigation just by looking at the leg from outside.

### Summary

In this chapter we have identified that the damage people have in their lower legs from venous reflux disease isn't from venous pressure (venous hypertension) but is due to an 'impact pressure' of refluxing venous blood - made up from the rate of change of pressure and the volume of the refluxing blood causing that pressure change.

We have seen that this combination of factors causes acute inflammation on standing in people with phase 2 passive reflux in their GSV and total incompetence of their deep veins; and that this inflammation starts deeply around the veins and capillaries and so cannot be assessed or even seen externally early in its development.

However, although we have an understanding of the venous reflux and how it causes damage via inflammation, this has not explained why some people with venous reflux get varicose veins and others do not. The next chapter will explain this.

# CHAPTER 9

# Varicose veins - the 'good guys'

When I was a junior surgeon, I attended a lecture by one of my heroes of surgery, Professor Andrew Nicolaides, who posed the following question: "Why is it that one person who has lost all of their valves in the Great Saphenous Vein will develop varicose veins and another person with the same underlying problem will develop a leg ulcer but no varicose veins?"

This question was one of the most formative moments in my life – it prompted my investigations into venous disease and has resulted in my whole approach to venous disease; culminating in this book and my approach to treating venous reflux disease.

Many people still ask me "Do varicose veins cause leg ulcers?" This question is a complete nonsense and cannot be answered with a "yes" or "no". By the end of this chapter you will not only understand why a simple "yes" or "no" cannot be given to this question, but you should also be able to explain what the correct answer actually is.

## Impact pressure

To understand venous disease, and particularly to teach it to others, I have always found a series of models to be useful.

To understand the impact pressure at the ankle from Phase 2 passive reflux in the Great Saphenous Vein, I usually use the model of either an exhaust pipe coming from an engine (for those who prefer cars) or a gun barrel coming from a pistol (for those who prefer guns or spy films) see **Figure 39**.

Whether considering an engine or a gun, the basic principle is that an explosion starts the process off. An explosion can be thought of as an exceptionally rapid expansion of gas from the substance that initially started as a liquid or solid. In a car engine a mixture of petrol and air is ignited within a cylinder, forcing a piston down

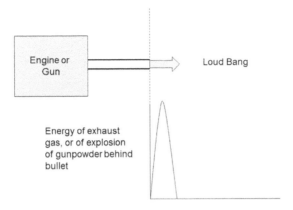

Figure 39: Model demonstrating high impact pressure

allowing the exhaust gases to be discharged into the exhaust pipe.

In a gun, the cartridge full of gunpowder explodes with the gas pushing the bullet along the barrel and, once the bullet leaves the barrel, the exhaust gas escapes from the end of the barrel.

In both of these instances, the surge of exhaust gas from the explosion bursts out of the end of the long, straight tube, either exhaust pipe or gun barrel.

Almost all of the molecules in the gas are travelling at high energy, meaning that when the gas hits the normal air, there is a huge impact pressure transferred in a very short period of time, which makes a violent shock wave in the air. This will be heard as a very loud bang and can be felt by any observer close enough.

**Reducing impact pressure with an 'silencer'**

As we are all aware in our normal day-to-day experience, cars are not that noisy any more. This is because they have a 'silencer' fitted. Those of us who watch spy films are aware that a good spy screws a 'silencer' on the end of a pistol before shooting it, so that the gunshot is not heard. So how does a silencer work?

This is explained in **Figure 40**. The explosive process is the same, whether it is an engine or gun. The difference is that on the end of the long tube (either the exhaust pipe or the gun barrel) there is an additional component added.

This additional component is basically another tube, which is far larger in diameter than the original. This allows the gas coming from the end of the tube to expand as if it were leaving the tube. Hence this is often called an 'expansion chamber'.

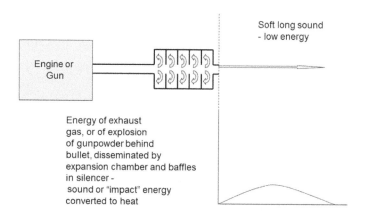

Figure 40: Model demonstrating how a high impact pressure can be reduced by a 'silencer'

In this expansion chamber, there are many metal discs, called baffles, with a hole through the middle. This segregates the expansion chamber into many different sections. At the other end of the expansion chamber, the tube (exhaust pipe or barrel) then continues at the same diameter as previously.

This combination of an expansion chamber containing baffles, inserted along the course of the exhaust pipe or barrel, is called a 'silencer'. It works simply by reducing the impact pressure of the gas being released after the initial explosion.

Working through the process, the explosion happens and the exhaust gas is released as a short and high-energy shock wave into the tube (the high energy gas molecules moving very fast). The gas then travels along the tube into the expansion chamber. At this point, most of the gas spreads outwards, starting to create the shockwave in the stationary air that it has hit.

However, this shockwave cannot spread out far, as the high energy, fast-moving gas molecules hitting the first baffle and the sidewall of the silencer (or expansion chamber) lose energy and speed in the collision. As energy cannot be destroyed or created, the energy of the shock wave is converted into heat in the metal baffle and chamber wall, caused by the impact of the gas molecules against the metal.

The escaping gas, now at lower energy and lower speed, and therefore with less impact pressure potential, spills through into the second chamber of the silencer where the same process is repeated. This continues throughout the length of the silencer until finally, when the escaping gas reaches the continuation of the tube, it has lost most of its energy and is moving relatively slowly. Here it escapes into the atmosphere as a long and low energy impulse with very little impact energy.

This very low energy impulse hardly affects the standing air at the end of the exhaust or barrel and so there is very little sound heard – i.e. the explosion has been 'silenced' and the impact energy of the gas has been lost by being converted into heat within the silencer.

## Impact pressure as a cause of venous disease

We can now take the same principles and apply them to the venous system (**Figure 41**).

As we have seen in Chapter 3, there are no functional valves in the inferior vena cava and pelvic veins (iliac veins) in the majority of people. In people with complete incompetence of the Great Saphenous Vein, blood travelling from the right atrium to the groin is able to continue its journey downwards due to gravity, from top to bottom of the whole of the Great Saphenous Vein. This is the phase 2 passive reflux in the Great Saphenous Vein that we discussed in Chapter 5.

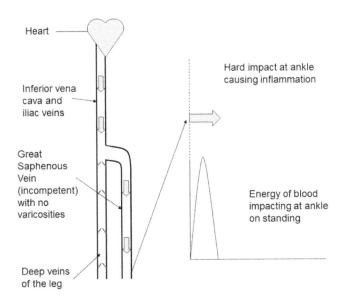

Figure 41: How a high impact pressure hits the ankles in phase 2 passive reflux of the GSV

If we now look at the ankle in a person with this condition, every time they stand up there is a sudden rise in impact of pressure from the refluxing blood which, as we explored in Chapter 8, causes inflammation in the tissues around the venules.

Although the inflammation from one such impact is very small, the effect accumulates over time. Over many years, this inflammation ends up causing the venous damage at the ankle as we have discussed.

### The role of varicose veins reducing the damage caused by impact pressure

Just as we use an expansion chamber with baffles to silence the impact pressure from the escaping exhaust gases from an explosion in a cylinder in an engine, or from the explosion of gunpowder pushing the bullet out of a gun barrel, the body tries to reduce the damage at the ankle due to the impact pressure of the column of blood falling from right atrium to ankle down an incompetent Great Saphenous Vein (**Figure 42**).

In the venous system, normal tributaries (veins) feeding into the

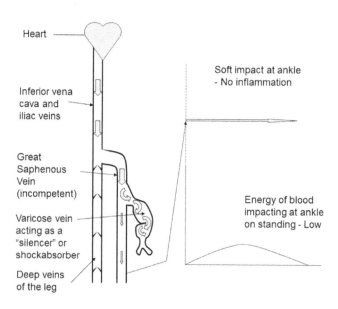

Figure 42: How varicose veins act as a 'silencer' to reduce the impact pressure and hence inflammation at the ankles in phase 2 passive reflux in the GSV

Great Saphenous Vein often become incompetent and dilate. This then allows the refluxing blood to be sidetracked from refluxing straight down the main Great Saphenous Vein and from hitting the small veins and venules at the ankle.

Indeed, some of the tributaries that become incompetent and dilate may well have been the original veins that became incompetent in the leg and might well have been those responsible for starting the phase 1 passive reflux at the beginning of the process (see Chapter 5 regarding ascending reflux).

If we look at the effects that such dilated tributaries might have on the venous reflux in the Great Saphenous Vein, we will start to understand the major point of this chapter.

We know that in the absence of these incompetent tributaries, the repetitive impact pressure of the phase 2 passive refluxing blood down the Great Saphenous Vein causes inflammation a the ankle. However, when this refluxing venous blood is sidetracked into these veins from the Great Saphenous Vein, the impact pressure causes further dilatation of the walls of the tributaries, stretching them and elongating them further.

Due to the elastic nature of the venous wall in these tributaries, the potential energy in the blood refluxing down the incompetent Great Saphenous Vein is changed into a combination of mechanical (kinetic) energy in the work done by stretching the vein wall and further kinetic energy losses through mild heating of the vein wall.

This process, when it occurs, has two major advantages for people with grade 2 passive venous reflux. The first is the impact pressure is directed away from the small veins and venules at the ankle, reducing inflammation and tissue damage there. The second is that the process of converting the potential energy of the impact pressure into kinetic energy of dilatation of the vein walls of the tributaries, as well as some heat energy, reduces inflammation and therefore damage to them or the surrounding tissues.

As these tributaries dilate and get bigger, they are much more likely to be seen on the surface and therefore be noticed and diagnosed as 'varicose veins'. It is essential to note at this point that any 'varicose veins' that are seen on the surface are only dilated tributaries and are not the route of the major reflux in this simple model.

This brings us to one of the very important points of understanding venous reflux disease and, in particular, the clinical diagnosis and the planning of treatment for varicose veins. The varicose veins that are visible on the surface and that everybody worries about (because they can see them), are actually beneficial to the body. They are transferring the energy of the impact pressure of the Phase 2 passive reflux in the Great Saphenous Vein to a harmless stretching of the wall, protecting the venules and capillaries at the ankle which otherwise would have had damaging inflammation.

Any treatment strategy that removes these varicose veins without taking away the underlying Phase 2 passive reflux is actually worsening the clinical situation, increasing the risk of venous damage at the ankle leading to tissue damage and eventual ulceration – despite making the patients leg look better because the lumps have been removed!

## The effect of varicose veins on the venous refill time in Great Saphenous Vein incompetence

As we described in Chapter 8, the phase 2 passive reflux in the Great Saphenous Vein is associated with a marked reduction in venous refill time when a patient stands up from lying down.

The presence of large varicose veins helps to keep the refill time longer. This reduces the rate of change of pressure at the ankle on standing and therefore reduces the inflammation at the ankle. The reduction of inflammation on standing reduces the risk of tissue and skin damage.

Therefore the bigger and more numerous the varicose veins, the bigger the reduction in refill time, the less the impact pressure at the ankle and the less the inflammation and skin damage at the ankle. This can be seen graphically in the **figure 43**.

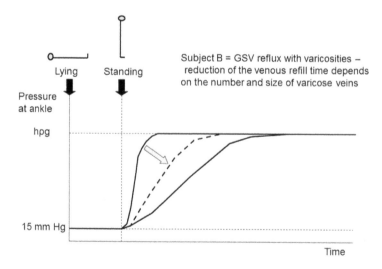

Figure 43: How varicose veins reduce the rate of change of pressure in the veins at ankle in phase 2 passive reflux of the GSV

### 'Varicose veins' and venous reflux or 'hidden varicose veins'

This is an ideal time to point out one of the difficulties in medicine: trying to use words that have been developed historically to describe a visible condition.

Early physicians would describe what they could see or feel. We have a much deeper understanding of the body and how it works, thanks to hundreds of years of research, surgery and imaging machines. Unfortunately though, the traditional names tend to live on and often serve to confuse.

An example of this is 'stroke'. When people had strokes over 1,000 years ago, physicians could see that one side of the body was affected. As it happened suddenly, the person was thought to have been 'struck down' by a deity for some reason, and so the name was appropriate. If the left arm, left leg and left face were affected, it was clearly a 'left stroke'.

Thanks to post-mortem studies, CT Scans, MRI Scans and a host of researchers, we have known for a very long time that the right brain controls the left side of the body, and the left brain controls the right side of the body.

As most strokes occur because of death of brain tissue due to a bleed or clot in the supplying blood vessels, the event causing the death of the brain tissue is called a cerebrovascular accident – or CVA. Hence a 'left stroke' is caused by a CVA on the right side of the brain. However when doctors communicate with the public or even each other, there is often confusion as to which side is being talked about and terms such as 'right brain stroke' are sometimes heard.

In the world of veins, a similar confusion is very widespread amongst doctors, nurses as well as non-medical people. When bulging and dilated veins are seen and/or felt on the legs, a diagnosis of varicose veins is made. We know now that these are not the problem, but are a sign of venous incompetence in deeper veins that we cannot see on the surface. But what about patients who have venous reflux but who do not have visible varicose veins? Research suggests that for everyone with visible varicose veins, there is another with venous incompetence and venous reflux disease who do not have visible varicose veins.

Thus, as you will now understand, the main problem that is caused by venous disease (as far as we have covered up to this point) is that of passive venous reflux in veins whose valves are not working. When this originates from the heart and continues down an incompetent Great Saphenous Vein (phase 2 passive reflux) and if no tributaries dilate, and therefore no varicose veins are seen on the surface, patients do not believe that they are suffering from 'varicose veins' despite the fact that they will be accumulating damage due to repeated inflammation at their ankles (see below).

For those fortunate people who have developed varicose veins, not only do they know they have a problem, but they can show doctors, nurses and other healthcare professionals a visible problem. If these professionals understand venous disease even to a small extent, they will refer that patient to a venous expert for diagnosis and treatment. Of course there are many who do not understand venous disease, still tell these poor patients that the visible veins are 'cosmetic only', and therefore allow them to deteriorate further by neglect.

Much more complex are the people with the phase 2 passive reflux who have not developed varicose veins and have worse inflammation at the ankle as a result, due to lack of the 'silencer' effect. Clinically these people often complain of aching legs on standing (particularly by the end of the day), swelling of the ankles, venous eczema, itching of the skin around the ankles, discolouration of the skin of the lower leg and ankle which is lipodermatosclerosis (a hardening of the skin and subcutaneous tissue of the lower leg frequently associated with a change in texture and colour) and eventual venous ulceration.

Because of the lack of understanding of venous disease, in both healthcare professionals as well as the public at large, such patients are often told that they do not have a venous problem and are often leave as 'palmed off' with support stockings or steroid/anti-inflammatory creams, neither of which get to the underlying cause of the problem and therefore only ever provide temporary and symptomatic relief.

Leg ulcers, in particular, are often only ever treated with bandaging or compression, and rarely seem to get referred for proper investigation which could lead to the correction of the underlying problem and cure. As patients rarely have an understanding of 'venous reflux' in the absence of visible varicose veins, they usually accept these temporary solutions as 'treatment' and do not push for proper diagnosis and treatment, which could well cure their problem.

One of the ways I have been trying to help the general public to understand this problem is to change my use of the terms 'venous incompetence' and 'venous reflux' in describing the underlying problem and instead use the term 'hidden varicose veins'.

Although 'hidden varicose veins' is technically incorrect, it does convey the basic understanding that there is a venous problem that is not visible on the surface and helps people link venous diseases together, despite their different clinical appearances. I would therefore commend the use of this term when communicating to non-medical audiences.

## Summary

By now you will have realised the nonsense of the question "do varicose veins cause leg ulcers?"

Phase 2 passive venous reflux in the Great Saphenous Vein can either cause varicose veins if the body responds by trying to prevent inflammation, or, if this does not occur, results in inflammation at the ankle which eventually results in leg ulcers, if the underlying venous reflux is not corrected.

Therefore the phase 2 passive venous reflux can cause varicose veins. It can also cause leg ulcers. However, varicose veins themselves do not cause leg ulcers (**Figure 44**).

Of course the two conditions can exist together. If the response of the body to dilate the tributaries to make varicose veins is insufficient to reduce the inflammation at the ankle, or if the reflux has continued to worsen as the Great Saphenous Vein dilates so that the damping effect of the varicose veins on the venous reflux impact pressure has become inadequate, then despite the appearance of varicose veins, inflammation at the ankle may start occurring which with time can deteriorate through the stages to venous ulceration.

Now we understand the problems with venous reflux disease, it is important to point out that all of the arguments so far have been looking at the simplest pattern of venous reflux - that of passive reflux down the Great Saphenous Vein.

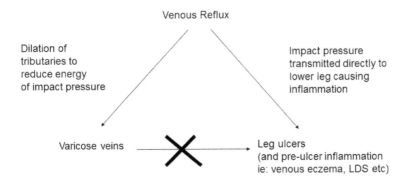

Figure 44: A common – and nonsensical – question: "Do varicose veins cause leg ulcers?"

In the next chapter, we will start exploring different patterns of venous reflux and in particular, active venous reflux, which has been misunderstood for many years leading to many erroneous statements and concepts made by clinicians previously thought to be 'expert' in venous surgery.

# CHAPTER 10

# Patterns of reflux - passive and active phase reflux

So far we have discussed passive phase reflux and have discovered how it is phase 2 passive reflux, with it's impact pressure from the height of the right atrium, that either causes dilatation of tributaries and visible varicose veins or the inflammatory changes at the ankle in people without varicose veins to absorb this energy.

In Chapter 5 we discussed the difference between phase 1 passive reflux of the Great Saphenous Vein, where the volume of blood is low and the height of the blood column is low thereby causing little in the way of inflammation at the ankle in most people, with smaller less tense varicose veins, and phase 2 passive reflux of the Great Saphenous Vein where far larger volume of blood refluxes from far higher creating far more energy in the impact pressure and therefore causing far more inflammation at the ankle.

However, this discussion so far has been restricted to the Great Saphenous Vein as this is the easiest to understand (**Figure 45**).

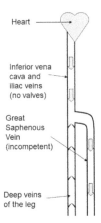

Figure 45: Patterns of reflux – phase 2 passive reflux in the GSV – due to gravity (and incompetent valves)

In real life, people with isolated Great Saphenous Vein reflux and no other venous problem are among a minority of patients presenting with venous problems, provided that sufficient time, energy and expertise is used to make a full diagnosis.

We will now look at other patterns of venous reflux that affect the lower limb. We will start off by looking at other patterns of passive phase reflux, as these are easier to understand with the principles we have discussed up to now.

## Other patterns of passive phase reflux – Pelvic Venous Incompetence

As we discussed in Chapter 5, there are no functioning valves in the veins draining the legs, above the groin (the pelvic (iliac) veins and inferior vena cava).

However these veins do not drain some of the pelvic organs, such as the ovaries or testicles, the uterus and the bladder. In addition the lower bowel also drains by another vein, but that does not form part of this discussion at present and so will be ignored for now.

We are going to think about the veins that drain the gonads – the ovaries in the female and the testicles in the male. In the embryo, the gonads develop high in the abdomen, close to the kidneys (an area called the 'urogenital ridge'). As the embryo develops into a foetus, and then into a baby, the gonads descend taking their blood vessels with them.

This process results in the gonads having long veins draining blood from them that ascend up the posterior part of the abdomen. These veins insert into the renal vein on the left (the drains the left kidney) and into the inferior vena cava on the right. Just as with the veins in the legs, these veins have to have valves in them to prevent blood from falling back downwards with gravity and causing varicose veins around the gonads.

Here we should discuss the differences between the male and female, so we can concentrate on the relevant area for venous reflux in the legs from the pelvis: venous reflux in females.

In males, the testicle descends through the pelvis and then out through the anterior abdominal wall, coming to rest in a sack of skin and muscle called the scrotum. If the valves in the testicular vein or veins become incompetent, passive reflux occurs backwards down the testicular vein, dilating the veins around the testicle in the scrotum. This condition is called a 'varicocele'. As the scrotum is on the outside it can be seen or felt easily (often described as being a 'bag of worms') and so varicoceles have been recognized for thousands of years.

As varicoeles are confined to the scrotum and do not connect downwards with veins in the legs, they do not play any further part in our discussions here. They do need treatment due to reasons of fertility and comfort but that is not the subject of this book.

In females the situation is far more complex. The right and left ovaries descend into the pelvis where they lie either side of the uterus. Each has a long draining vein: the ovarian vein. However, as they are inside the body and are closely related to other pelvic organs, some of their blood also drains via other pelvic veins (the internal iliac veins). We will not discuss this further here, but it will become important when we discuss treatments in another book in this series.

When the ovarian veins become incompetent, the veins dilate around the ovaries causing 'ovarian varicoceles'. Of course these are deep inside the pelvis and so are not visible from the outside. As such they have been (and still are) ignored by the majority of doctors and nurses.

These varicose veins inside the pelvis dilate, pressing on other structures in the pelvis. If small, they may cause no discernable symptoms. However, if they enlarge, they may cause a 'dragging' feeling due to the weight of the blood in the dilated pelvic veins resting on the pelvic floor, particularly when they dilate pre-menstrually; or they may push on the bowel (causing irritable bowel), bladder (irritable bladder) or vagina (discomfort on sexual intercourse). Any of these symptoms can occur by themselves or in combination. When present this is called 'pelvic congestion syndrome'.

It is a sad reflection on medicine today that despite there being so many wonderful advances in many areas, there are huge numbers of women suffering from pelvic congestion syndrome who never get a diagnosis. Family doctors and gynaecologists when confronted by someone with these symptoms usually check for cysts, tumours, endometriosis and adhesions amongst many conditions that they are trained to recognise but few look for the dilated veins of pelvic congestion syndrome.

Turning to the relevance of pelvic venous reflux to this book, it is important to note that this reflux is a passive reflux caused by gravity and the failure of valves in the ovarian veins. The resulting reflux tends to cause varicose veins in the pelvis (ovarian varicoceles) but these veins can connect with veins in the legs.

The usual communication is via veins in the wall of the vagina, although other routes of communication exist. However in women who have not been pregnant, these communicating veins usually have competent valves and so refluxing blood cannot pass into the legs. As with most things in medicine, there are exceptions and we have seen a few women who do have incompetent vaginal veins despite having never been pregnant.

It appears that when some women who have pelvic venous reflux have a pregnancy and then a normal vaginal delivery, the vaginal veins can become incompetent. The exact cause isn't known as this whole cause of venous reflux in the legs is still a relatively new discovery. However, just from logic, it is probably due to the trauma of the immense stretching that is required of these veins during the birth process.

In women with reflux in their pelvic veins who go on to get incompetent vaginal veins, blood can now reflux passively from the right atrium, down the upper inferior vena cava, down the ovarian veins, through the vaginal veins and into veins of the thigh where

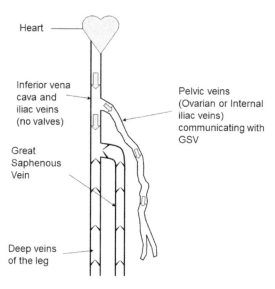

Figure 46: Patterns of reflux – passive reflux in the pelvic/ovarian veins – due to gravity (and incompetent valves)

they can cause the veins to dilate and become varicose (**Figure 46**).

In some women, these veins can communicate with the Great Saphenous Vein, causing, or adding to, reflux in this main vein (**Figure 47**).

Since we started popularizing this problem in 2000 on our website www.vulval-varicose-veins.co.uk, we have found that this route of reflux is usually missed by traditional varicose vein surgeons or doctors, and this accounts for very large numbers of women who keep getting their veins back after treatment. Traditional teaching shows that all reflux comes from the leg veins – hence pelvic and

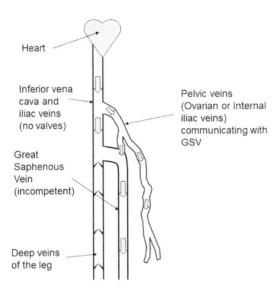

Heart

Inferior vena cava and iliac veins (no valves)

Pelvic veins (Ovarian or Internal iliac veins) communicating with GSV

Great Saphenous Vein (incompetent)

Deep veins of the leg

Figure 47: Patterns of reflux – passive reflux in the pelvic/ovarian veins – joining and adding to passive reflux the incompetent GSV in the leg

ovarian veins are rarely checked nor treated. As such these women don't really have recurrent varicose veins, they simply have never had the correct investigations or treatment in the first place.

One final note about the passive reflux from the pelvic veins: this reflux does cause varicose veins in the vulva, upper thigh and leg (especially in the inner thigh and back of the thigh). However, as the veins have a relatively small diameter compared with the inferior vena cava and iliac veins, and as they meander through the pelvis rather than run straight downwards, the refluxing blood is of low volume and has low impact pressure. The potential energy of the refluxing blood is lost as it impacts on the walls of multiple tortuous veins in the pelvis and vagina, meaning that this reflux virtually never causes any inflammatory problems at the ankle.

## Active Phase Reflux – The Small Saphenous Vein and Incompetent Perforating Veins

We have now been through so much understanding of the venous system of the legs, that you are probably feeling quite confident in your understanding of venous reflux, varicose veins and the causes of the venous skin damage at the ankles.

When I lecture on this subject, this is the point where I like to throw a spanner into your understanding, to make you think around the subject and to show that venous reflux disease is not as simple as many think – and is certainly not as simple as it is taught traditionally.

For this argument, consider the Small Saphenous Vein (**Figure 48**).

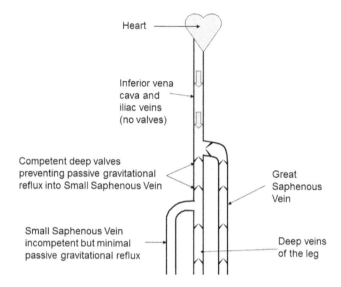

Figure 48: Patterns of reflux – the incompetent Small Saphenous Vein, how does it reflux?

Clinically we know that patients with significant reflux in the Small Saphenous Vein go on to get the same problems as people with significant reflux in the Great Saphenous Vein: varicose veins or inflammatory changes (swelling, venous eczema, lipodermatosclerosis, leg ulceration etc). My question is how does this happen?

Using our knowledge of phase 1 and phase 2 passive reflux, we can see that the valves in the deep vein above the Small Saphenous Vein junction will stop passive reflux from the heart, making Small Saphenous Vein reflux the same as Phase 1 reflux only. The height of the column of blood refluxing down the Small Saphenous Vein is only the height of the knee.

Even though the diameter can be big, giving a large volume of blood to reflux, the lack of height of the column means that the impact pressure is very low. This observation has led some surgeons who do not understand venous disease to suggest that the incompetent Small Saphenous Vein doesn't need treating.

The argument is compounded further when we think about Incompetent Perforating Veins (**Figure 49**). These small diameter veins are often below the knee and, when incompetent, have low volumes and are protected by more valves in the deep veins,

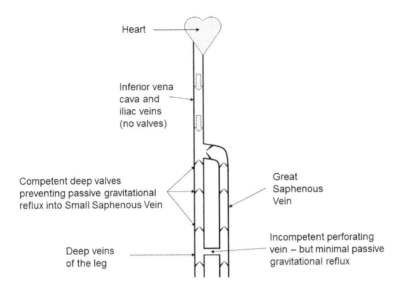

Figure 49: Patterns of reflux – the incompetent perforating veins –
how do they reflux?

making the height of any column of blood negligible. Since the
impact pressure from passive reflux is negligible, many researchers
and famous vein 'experts' have recommended that Incompetent
Perforating Veins do not need treatment and can safely be left if the
Great Saphenous Vein reflux is treated adequately.

However this flies in the face of experience when we find patients
with varicose veins, or venous damage to the ankle, who only have
reflux in the Small Saphenous Vein or incompetent perforating veins
(and that these changes reverse and are cured on successful treatment
of this efflux).

So, the question is "how" does this reflux cause the problem, and
not whether it does or not.

## The Venous Pump and Active Phase Reflux

The error in the thinking of those who do not understand this
problem is fairly easy to understand. Once doctors and nurses
understand passive reflux, they look for the same mechanism
everywhere they see varicose veins or the squelae of venous damage
at the ankle.

Rather than looking for 'more of the same', which clearly doesn't
exist in the incompetent Small Saphenous Vein and incompetent
perforating veins they should be looking for another mechanism of
venous impact pressure.

This second mechanism which is responsible for the effects of the venous reflux found in these difficult to understand veins, is one of active phase reflux.

Active phase reflux is a result of the combination of the venous pumps in the leg working, when some of the veins coming off the side of the main deep veins have valves which are not working.

In my lectures, I find that understanding comes quickest and easiest if I use a model **(Figure 50)**.

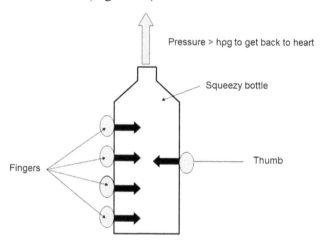

Figure 50: Model of the venous pump in the leg

Think of a plastic 'squeezy' bottle, such as those used to hold detergents or washing up liquid in the household. This is our model of the deep veins. We need to pump blood out of the deep veins and up to a height of about 1.5 meters in a normally sized adult (i.e. enough pressure to get the blood up to the heart). As blood is more dense and viscous than water, you could think of having to pump water even further (maybe to about 2 metres).

Now to pump the fluid, you put your hand around the bottle ready to squeeze. Your hand is the calf pump in this model. Imagine squeezing hard enough to shoot the water 2 meters in the air. This is the action of the calf pump.

In order to understand active phase reflux in the Small Saphenous Vein and incompetent perforating veins, you are going to have to imagine repeating this action but having made a hole between your fingers. This allows water to 'actively reflux' during the 'pumping' **(Figure 51)**.

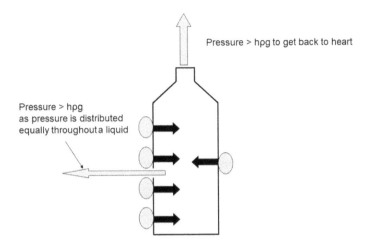

Pressure > hpg to get back to heart

Pressure > hpg
as pressure is distributed
equally throughout a liquid

Figure 51: Model of the venous pump in the leg with a 'hole' in the pump –
understanding active reflux as seen in SSV and IPV incompetence

It is easy to imagine the effect. As we learned in science at school,
pressure in a liquid is distributed equally in all directions. If enough
pressure is exerted by the hand to pump water straight upwards 2
meters against gravity, then the same pressure forces blood out of the
hole at very high speed and energy.

We can now relate this to the venous system (**Figure 52**).

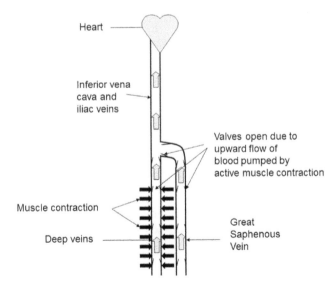

Heart

Inferior vena
cava and
iliac veins

Valves open due to
upward flow of
blood pumped by
active muscle contraction

Muscle contraction

Deep veins

Great
Saphenous
Vein

Figure 52: Activation of the venous pump in the leg - during muscle contraction
blood is pumped back to the heart through normal veins

When the venous pump activates in the leg, the deep veins are squashed under great pressure, putting the blood within the veins under great pressure and forcing it to shoot up the vein towards the heart.

However, just like in the plastic 'squeezy' bottle with a hole in the side, if the Small Saphenous Vein is incompetent, or any perforators are incompetent, this pressurized blood escapes out into the incompetent vein at high speed, causing an impact pressure wave down the vein (**Figure 53**). This wave of refluxing blood with high impact pressure acts in exactly the same way as phase 2 passive reflux does in the incompetent Great Saphenous Vein – it either hits incompetent tributaries and causes varicose veins, or passes down the veins hitting the venules, causing inflammation and tissue damage.

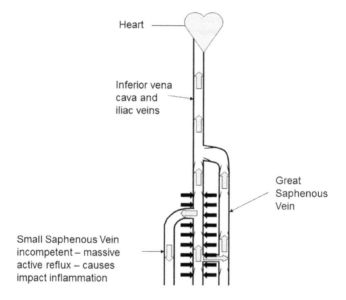

Figure 53: Active phase reflux - during muscle contraction blood under pressure in the deep veins is pumped at pressure out of the incompetent SSV and IPV

### What is the difference between phase 2 passive phase reflux and active phase reflux?

The the end results of phase 2 passive reflux down the Great Saphenous Vein and of active phase reflux of blood down an incompetent Small Saphenous Vein or incompetent perforating vein is the same. To the tributary being stretched or the venules being hit and causing inflammation, the impact pressure of the blood hitting them and the way the energy is dealt with is the same.

However the two different causes of this damaging venous reflux are markedly different when it comes to investigating and treating the patient.

Firstly, virtually all venous tests for varicose veins and venous reflux disease have been developed to measure or detect passive phase reflux. Although beyond the scope of this book, and discussed more fully in another book in this series, Doppler, duplex ultrasound and plethysmography tests traditionally rely on pumping the leg first, and then testing for any refluxing blood.

Such tests will clearly pick up the phase 2 passive reflux in the incompetent Great Saphenous Vein and its tributaries, passive reflux in the pelvic veins emerging from the top of the thigh and phase 1 passive reflux in the vein arising from incompetent pelvic veins.

However it is likely to miss, or massively underestimate, any active phase reflux in the Small Saphenous Veins or perforators. Simply, the measuring part of these test are too late to pick up the damaging reflux, which occurred during the pumping phase of the test. They 'missed the boat' and are therefore useless in their current form.

Secondly, as active phase reflux is not well understood by most doctors and nurses – even those treating patients with varicose veins, venous leg ulcers and other venous reflux problems – the fact that their tests don't show the active phase reflux often leads them to incorrectly overlooking the need to treat reflux in the Small Saphenous Vein or in incompetent perforating veins.

Thus the venous literature for the last hundred years is full of 'experts' claiming that incompetent perforating veins can safely left untreated and even some who suggest the same for an incompetent small saphenous.

I hope that readers of this book will join those who have attended my lectures and teaching courses, realising the importance of finding all sources of reflux in the leg veins, and treating them with appropriate methods.

# CHAPTER 11

# Conclusion

This book is a summary of my work in this area over the last 20 years of my life. It represents my understanding of venous reflux and is made up of countless hours of attending conferences, reading research papers, performing research and, of course, investigating and operating on my own patients.

I have provided an overview of the subject, trying to make the concepts easy to understand and fun to learn. In this way I hope that anyone reading it will realize the many failings that exist in the world of venous investigations, surgery and other treatments; and and will understand that they can be rectified.

My belief is that anyone treating venous reflux disease, whether it be cosmetic treatment of thread veins (which usually has an underlying venous reflux feeding into the area of concern), varicose veins, venous leg ulcers or any of the other conditions caused by venous reflux disease, should have a thorough understanding of everything within this book.

As venous reflux disease affects between 1 in 2 and 1 in 3 of the population, it would also be beneficial if interested members of the public also knew the story inside this book, so they could start to identify venous reflux disease in themselves or their loved ones; and could also be aware of the steps that their chosen venous 'specialist' should take in investigating them prior to treatment.

In order to make this text short and readable, I have not attempted to reference the hundreds (if not thousands) of talks, articles, papers and books that have been amalgamated with my own extensive experience of venous reflux, that makes up my understanding of this fascinating subject.

Lightning Source UK Ltd.
Milton Keynes UK
UKHW020953140622
404410UK00006B/302